The ITDG Occasional Papers Series

1. Marilyn Carr, *Appropriate Technology and Rural Industrialisation.*
2. Derek Miles, *Appropriate Technology for Rural Development: The ITDG Experience.*
3. John Davis, *New Directions for Technology: The Tawney Lecture 1981.*
4. George McRobie and Marilyn Carr, *Mass Production or Production by the Masses?*
5. Marilyn Carr, *Women and Appropriate Technology: Two Essays.*
6. Alan Bollard, *Industrial Employment through Appropriate Technology.*
7. Richard Whitcombe and Marilyn Carr, *Appropriate Technology Institutions: A Review.*
8. Sanjay Sinha, *Planning for Rural Industrialisation: A review of developing country programmes.*
9. Frances Sinha, *Women, Work and Technology in Rural South Asia.*
10. Marilyn Carr, *Intermediate Technology in Botswana: A Review of the Botswana Technology Centre.*
11. Stephen Biggs and Ruth Grosvenor-Alsop, *Developing Technologies for the Rural Poor.*
12. Marilyn Carr, *Intermediate Technology in Papua New Guinea: A Review of SPATF.*

Intermediate Technology publications are available from the I T Bookshop at 9 King Street, Covent Garden, London WC2E 8HN, UK, or by mail order from the same address.

Developing Technologies for the Rural Poor

Stephen Biggs and
Ruth Grosvenor-Alsop

Practical Action Publishing Ltd
27a Albert Street, Rugby, CV21 2SG, Warwickshire, UK
www.practicalactionpublishing.org

First published 1984 \Digitised 2013

Reprinted by Practical Action Publishing
Rugby, Warwickshire UK

ISBN 10: 1 85339 365 7
ISBN 13: 9781853393655

ISBN Library Ebook: 9781780441986
Book DOI: http://dx.doi.org/10.3362/9781780441986

A catalogue record for this book is available from the British Library.

Since 1974, Practical Action Publishing has published and disseminated books and information in support of international development work throughout the world. Practical Action Publishing is a trading name of Practical Action Publishing Ltd (Company Reg. No. 1159018), the wholly owned publishing company of Practical Action. Practical Action Publishing trades only in support of its parent charity objectives and any profits are covenanted back to Practical Action (Charity Reg. No. 247257, Group VAT Registration No. 880 9924 76).

CONTENTS

I. INTRODUCTION

This paper is a selective review of case studies of where government and non-government organisations (NGOs) have been involved in rural/agricultural technology programmes specifically directed at benefiting the rural poor. The purpose of the review is to try to identify specific institutional features which characterise organisations and agencies which appear to have benefited poor client groups in the short and long run. It is not a 'state-of-the-art' review and does not claim to give a representative coverage of relevant literature. Rather, it is a presentation of case study material aimed at illustrating and supporting what are felt to be the important institutional issues concerning the generation and diffusion of rural/agricultural technology.

The report is structured into five main technology areas, namely: crops, irrigation, post-harvest, draught animal technologies and livestock. Within each of these, case studies are divided into different types of institution, namely: universities and research organisations, government agricultural departments, non-governmental organisations, international development agencies, and the 'informal' sector. In the last category, situations are described where individual local artisans and farmers have, through purposive selection, trial and error, developed useful technologies without the major involvement of a formal agency.

As one might expect, the distinction between types of agencies becomes blurred at the edges. In addition, there is a very real problem of how to characterise and draw lessons from situations where two types of agencies interact. For example, it is open to subjective judgement as to whether the recent development of 'successful' types of on-farm research methods is seen as primarily a result of the inputs of international agricultural research institutes, or mainly as the result of the help and other inputs given to international scientists by local scientists in developing countries. The implications of this type of analysis are important as they can affect where and how funds might be best allocated in the future.

In each case study, the focus is on lessons which relate to important issues. This could not have been done universally across the board because, quite frequently, the required information was not available.

Further, some of the information must be treated with caution, as, understandably, agencies often publish information which portrays themselves in a favourable light. The issues looked at include: the identification of intended poor client groups; the dynamic process by which agencies went about designing, implementing, monitoring and changing activities over time; the short term effects of agency programmes on the intended client groups and on other poor people; and the long term effects on strengthening local poverty focused research and extension capabilities.

The paper is concerned with looking at those programmes where the agency has defined its clients as some group of poor people. These may be poor small farmers, poor rural women, malnourished children, landless labourers, etc. Some agencies call these programmes 'target' group programmes. However, it is preferable to think of 'client' groups rather than 'target' groups. The term 'client' is a more neutral professional term which does not carry with it some of the top-down, elitist connotations sometimes associated with the 'target' approach. For example, when an agency is targeting resources at the poor there is often an implicit assertion that 'we' know what your problems are, and now we are delivering the solution.

As so many poverty programmes have missed their 'target' it is clear that agencies should be less confident that they are able to correctly diagnose the problems of the poor and effectively deliver the goods and services needed.

This is more than just a matter of semantics. It reflects a whole way of thinking about technology generation and diffusion. The conclusions that project staff drew at the end of the first year of the Caqueza Project illustrate a change from a target approach to a more humble client approach:

> . . . field work and the increased contact with farmers allowed the project staff to identify several unforeseen areas of activity that, if neglected, appeared likely to substantially limit the

project's progress. Given these circumstances, they requested a substantial increase in staff for 1972. The old extension approach that considered the communication of the new technology to farmers as the only activity required was being forgotten and being replaced by the idea that more had to be known about the farmers' present production system before anything could be done about changing it. But agronomic knowledge alone was not enough; socio-economic knowledge was required as well. This was a year of observing the requirements for rural development to occur. The project staff began to comprehend that no surefire methodology existed, and that a long process of trial and error lay ahead of them.[1]

Sections II to VI of the paper contain the case studies by major technology area. Each case study ends with a summary of the major issues raised. General conclusions derived from the review are presented in Section VII.

II. CROPS

A. UNIVERSITIES AND RESEARCH ORGANISATIONS

1. CIMMYT (International Maize and Wheat Improvement Centre), Mexico and Kenya.

Kenya's 4th Five-Year Development Plan contains the observation that:

> Research must be of increasing relevance to the farmer's situation. This includes not only the physical environment that confronts him, but also the socio-economic setting of his farm activities.[2]

This statement reflects the concern that CIMMYT had expressed three years earlier when they made their commitment with the Kenyan Agricultural Research Services to Farming Systems Research (FSR). Although not restricted to poor small farmer situations, applications of FSR had been applied primarily to the problems of small farmers. Briefly, the FSR perspective is one that embodies the following characteristics:

(1) Farming systems research views the farm or production unit and the rural household or consumption unit – which in the case of small farmers are often synonymous – in a comprehensive manner. FSR also recognises the interdependencies and inter-relationships between the natural and human environments. The research process devotes explicit attention to the goals of the whole farm/rural household and the constraints on the achievement of these goals.

(2) Priorities for research reflect the holistic perspective of the whole farm/rural household and the natural and human environments.

(3) Research on a sub-system can be considered part of the FSR process if the connections with other sub-systems are recognised and accounted for.

(4) Farming systems research is evaluated in terms of individual sub-systems and the farming system as a whole.[3]

CIMMYT is thus concerned with the development of a technology[4] (a combination of all management practices for producing or storing a given crop or crop mixture) which is (a) appropriate to the circumstances of the farmer client group, and (b) helps to meet the national policy goals of the government. Therefore it attempts to reconcile local and national concerns to enable planning of effective research and development programmes.

Four collaborative regional programmes promoting FSR procedures have been established since 1976 with the funding of UNDP. The Eastern African Economics Programme was initially to focus on Kenya, Ethiopia, Uganda, Tanzania, Zambia and Malawi. Two examples of FSR in Kenya demonstrate the processes of this approach and problems revealed by it:

(1) Exploratory surveys of farmers growing intercropped maize and beans in Eastern Kenya threw new light on the interpretation of experiments in alternative mixture patterns. The surveys identified an acute labour shortage during crop establishment and showed that returns to labour required to establish the crop mixture would be a key criterion in appraising experimental results.

Recommended planting patterns for maize/bean mixtures require five times more planting labour than the simulated farmer pattern which gave almost four times the return to the planting labour used. In the farming systems of target groups growing maize/bean mixtures, which have a short rainy season and where land is not limited, there are often intense labour peaks at the time of crop establishment. For such client groups, return to seasonal peak labour used may be a more appropriate criterion than return per unit area in comparing results from experimental treatments.

(2) Exploratory and Verification Survey work was carried out in an area of Western Kenya with high population density and an acute scarcity of land. The results revealed a marked interaction between crop and livestock enterprises in the use of crop residues as byproducts for feeding local animals kept for milk . . . The dominance of maize as a source of cattle feed,

both green and dry material, led to proposals for two adaptive experimental programmes which were designed to examine:

(a) What increase in maize plant population would be possible so that fodder production could increase without penalising grain yields, in both the long and short rains, and

(b) The effects of alternative timing of picking the leaves and tops of maize on grain and fodder yields.

The second major leg of programme strategy was to build up the credibility of the Farm Economist with technical researchers, particularly agronomists. Here the programme has had limited success. Many of the problems encountered in establishing a close working relationship were features of the research organisation, particularly the strong compartmentalisation, upheld by everything from disciplinary loyalty to parallel compartmentalisation in the layout of government estimates and fund votes.[5]

Several important issues are exposed here.

There is a need to place a technological solution in the production environment for which it has been designed, i.e. farm level research is vital in the development and evaluation of the appropriateness of a technology. From the first example researchers learned that labour constraints were as much a pertinent issue to the cropping patterns as the physiological mixture. The logical progression in the articulation of this approach is then that the group for whom the technology is being developed must be clearly identified (a) to determine what is needed, and (b) to enable an evaluation of the impact of the proposed technology on that particular group. A clear definition of the intended client and beneficiary group will also assist researchers and field workers by giving them a point of reference in the monitoring of programmes.

In Zambia it was the small farmer that administrators identified as the client group for CIMMYT's FSR procedures. Partly due to the institutional problems experienced in Kenya a two level hierarchy – Commodity Research Teams and Adaptive Research Teams – was established, trying to train people from the start in this interdisciplinary approach.

This acknowledgement of the interaction of the different variables in a farming system highlights the structural and institutional problems of FSR, such as:

(a) the apparent difficulty of introducing the economist into an area previously dominated by the technical scientist, and

(b) the increasing emphasis placed on the roles played by the agronomist and the economist, who were 'perceived traditionally as playing service roles to disciplinary researchers' (Collinson, 1982). In the words of the same author 'The establishment feels threatened and the social scientist, seen as the intruder, is rejected.'[6]

There have been attempts to establish FSR procedures throughout East, Central and Southern Africa. However, there generally appears to be a lack of institutional 'acceptance' of the methodology, i.e. that FSR can reveal key areas previously undetected and of importance to the kind of technology that the technical scientist is developing for a specific client group. In concluding his analysis of FSR in Africa Collinson states:

> Lessons have also been learned from working with national research services. The most important is the recognition of the need for a flexible and pragmatic approach to different institutional situations and to the personalities involved in each situation. A major strategy is to focus on research services where there is already a strong awareness that research relevance is a problem. Within such establishments, if authority is strong, it may be helpful to introduce FSR procedures. Where direction is weak or conservative or where organisation is poor, new procedures can be seen as an added source of confusion – a nuisance. In such circumstances, only a bottom-up approach, working through the station and with individual scientists, seems feasible. Ideally, top-down authority and a bottom-up approach working through individual researchers can be complementary.
>
> A clear distinction has emerged between technical and adaptive research. Technical research is the solution of technical problems on research stations organised along disciplinary and commodity lines. Adaptive research is a selection and testing, from the range of potentially relevant technical solutions, of a partial or whole solution to a particular problem that has been established as a priority by a target group of farmers. A revised implementation strategy then is to establish adaptive

on-farm research teams, whose members build up their experience together, drawing on both the existing body of knowledge and on older disciplinary oriented specialists for potential solutions to identified systems problems. Once established, adaptive teams begin to channel unsolved technical problems back to the specialists. This process continues until problems identified on farms preoccupy both adaptive and technical researcher in the research hierarchy.[7]

A programme to develop on-farm research methods with a farming systems perspective (OFR/FSR) evolved in Mexico in the mid-late 1960s.[8] The programme was stimulated by the findings that although new seed varieties and practices were known to give higher yields the rate of adoption among farmers was actually very low (Perrin and Winkelmann, 1976). Partly motivated by this fact and partly by a recognition that all over the world the small farmer was suffering from, if not decreasing, then at least static, living standards, the Rockefeller Foundation in collaboration with CIMMYT and Mexico's Graduate School of Agriculture, set out to design a new programme. Plan Puebla, conceived of as a 'demonstration' rather than 'research' project (Redclift, n. d.) set out to solve the development problems – food shortages and low income in agriculture. However, although these problems were to some extent solved, the conception of the project as being outside the parameters of 'research' had to change. Institutionally it became undeniably obvious that successful technological adoption was unlikely to take place without some adaptive research. That this was done informally by the farmers themselves articulates a lesson that researchers and programme planners cannot ignore.

The client group of Plan Puebla was the traditional, resource-poor peasant or smallholder; the technology was to 'obtain massive increases in yield of the basic crop' (maize) (CIMMYT, 1969). The organisation featured coordinated efforts in agronomy, communications and evaluation, and the methodology, bearing the hallmark of FSR involved:

> . . . research in the farmers' fields, diffusion of technology and inputs through groups of farmers, continuing evaluation and feedback to the professional staff, coordination of the interests of farmers, plan staff and local institutions.[9]

(Redclift, n.d.) has analysed the project fairly objectively and it is probably best to leave it to his words to describe the 'facts' about Plan Puebla:

The project was conceived as a regional exercise, based on thirty-two *municipios* in the east of the State of Puebla, not far from the Agricultural University of Chapingo, which, with CIMMYT, took responsibility for the Plan. These municipalities comprised 116,800 ha. of cultivable land, of which 80,000 ha. was devoted to maize. In 1967 this 'region' included over forty-three thousand families of agriculturalists, a total population of almost a quarter of a million people. Of these *campesino* families, 38 per cent were *ejidatarios* (that is, they had private access to communal *ejido* land). Another 28 per cent were independent small producers and owners of the plots they worked. Average farm size was 2.7 ha., but the distribution of land within the project area meant that almost seventy per cent of the *campesino* families has less than two and a half ha. each.

The location for the project was chosen after consideration had been given to a number of factors. In the project area rainfall was relatively reliable, frosts were light and only occurred during the first quarter of the growing season, and soils were deep, permeable, and free from toxic amounts of salts. This physical environment was chosen on the grounds that favourable ecological growing conditions would permit the scientists to test the hypothesis that yields could be increased using existing technologies. The agronomic 'package' would consist of chemical fertilisers, 'improved' seeds, higher plant densities and other improvements in methods of cultivation. Had the project been located in an area that was ecologically more precarious, it would be less easy to demonstrate what could be achieved under reasonably favourable growing conditions. It is important to emphasise that Plan Puebla was not located in an area which could serve as an example of 'average' growing conditions, although it was often argued that the lessons from the project has very wide application in rain-fed agriculture

Another important consideration was the capacity of the *campesino* families involved in the project. The number of participants in the project rose from only thirty heads of families in 1967 to nearly five thousand in 1970. Of these almost eighty per cent were literate, a much higher percentage than in Mexico as a whole. Further, most of the *campesinos* who cooperated with the Plan were already well informed about the components of the technological package that was being offered them. Although project publications rarely refer to this fact the majority of participants in the project knew of the new synthetic maize varieties, and had used chemical fertilisers on their crops in 1967. Finally, a very favourable political and administrative environment existed in the State of Puebla at that time: the State Governor was enthusiastic, the project was situated in the home state of the Republic's President, and Leopoldo Solfs, the chief economic adviser to the President, was also enthusiastic about the Plan.

Discussion of results as reflected in the rate of adoption by farmers of the plans and recommendations are complicated by three issues:

(a) there was an initial tendency for farmers to only partially adopt an individual production practice. For example, instead of changing from a traditional level of 30,000 plants/ha to a recommended level of 50,000, the farmers often changed to some intermediate level; (b) farmers often tended to accept some recommended practices more readily than others: that is, they might increase their rate of nitrogen fertilisation before changing the way they apply it: and (c) farmers often tended to use the new technology initially on only a portion of their land.[10]

These issues highlighted the point that it was not possible to transfer complete packages of recommendations to the farmer. Another interesting point was in the selective use of parts of the technological package – farmers had discovered that local maize varieties were as responsive to fertiliser inputs as 'improved' varieties, and had achieved impressive increases in yield.

In brief summary then, can it be concluded that Plan Puebla was successful in reaching the client group? That maize yields of small farmers were increased must be acknowledged, but what is really interesting in terms of institutional lessons is threefold:[11] Firstly, the Plan demonstrated that it was possible to increase the production of the client group without recourse to massive investment. Secondly, through taking research into the farmers' fields, results were obtained that, although successful in terms of programme goals, were not those predicted. Thirdly it was learned that the credit policies for the small producer had to relate to his needs, resources and risk-averting behaviour (CIMMYT, 1974).

However, while some important lessons about the processes of technology generation and diffusion were highlighted in the Plan Puebla experience, other lessons concerning the strengthening of a sustained local research capability were not analysed in depth. Plan Peubla was a 'special project'. It had special inputs of staff (some expatriate), of research resources, of political patronage etc. The problems CIMMYT has faced in transferring, developing and trying to institutionalise the concepts of on-farm research with a farming systems perspective (OFR/FSR) have arisen because of the various socio-economic, political and resource environments in which scientists work in different developing countries. The problems on non-adoption of FSR in Africa, described by Collinson (1982) well illustrate that more attention needs to be given to an understanding of the local organisation and management of agricultural science before the principles of OFR/FSR can be strengthened in different professional environments. In cases where OFR/FSR appear to have worked well, the following lessons appear to hold:

(a) Experimental station and on-farm research programmes are under the same management.

(b) Constructive conflicts (Rhoades and Booth, 1982) exist most often when different disciplines have very sharp arguments as to what is important and relevant in different situations.

(c) Farmers' adoption of technology is a major criterion of 'success' for the programme.

(d) Peer group review and arguments amongst scientists within the same disciplines must be maintained. This peer group review has to be in the context of the specific field problems being faced by farmers – not in the context of academic journals etc.

(e) Local agricultural and social scientists who are able to evolve agricultural research and extension systems which are viable and sustainable in a local situation are (implicitly or explicitly) highly capable and sensitive to cultural, social and political influences which affect research.

(f) A strong research and extension system is dynamic and exhibits an ability to change, often very quickly and significantly (Biggs, 1982).

Needless to say, we find that some 'successful' research systems in the past have had these characteristics. For example the Adaptive Research Programme in Tamil Nadu, India in the early 1960s (Gaikwad, 1977). It is important to note that the current interest in OFR/FSR reflects to some extent a return to, or the filling out of, important principles found in the early chapters of most elementary text books on agricultural research and Survey methods. It is also important at this stage that the proponents of FSR start putting their methods and principles into a broader policy framework if their efforts are to benefit poor farmers.

In the following sections we look at other recent examples of where different forms of action research, on-farm and farming system research have been conducted in developing countries. Much of the discussion is on the methods and techniques used as few writers have analysed the institutional determinants of what has led to or undermined the strengthening of a local capability.

The influence of credit

Credit has been one of the issues that CIMMYT has recently looked at as regards the diffusion of improved cereal varieties. They have to date been involved in a number of studies examining the reasons why small farmers 'lag behind' large farmers in the adoption of technology. A number of variables have been identified, such as tenurial arrangements and the risk-aversion behaviour of subsistence or marginal farmers. The results of this analysis have sometimes led to credit being labelled as the production factor that inhibits adoption. It was not the purpose of CIMMYT to look in any depth at the pros and cons of policies emerging from such problem identification, rather its aim was to took at technology adoption constraints. Perrin and Winkelmann (1976) assessed six of CIMMYT's adoption studies and found that farmer behaviour was not always related to those variables mentioned above, but did have some correlation with the use or availability of credit. However, the existence of credit schemes was not a deciding factor in adoption decisions. What they found was that technology would be adopted if it were profitable. If a formal credit scheme were available then this would be used, but if not then money for inputs would be borrowed elsewhere. By the same token, Brammer (1978) found evidence in Western areas of Bangladesh that farmers were 'paying open-market prices up to two to four times the official price for HYV wheat seeds'.[12] The new wheat seeds were very appropriate and very profitable. In some technology generation and diffusion programmes, the lack of credit, inadequate prices and unfavourable tenure conditions are blamed for low rates of technology adoption. While recognising that this may be the case in some situations, we should recognise that many research and diffusion programmes have not produced viable technology because of the management and organisation of the programmes.

In conclusion, Perrin and Winkelmann (1976) state:

> The adoption studies with which CIMMYT has been associated have shown that to a limited extent differences in farmer adoption behaviour can be explained by differences in information, in the availability of inputs, in market opportunities for the crop, and differences in farm size and farmer risk aversion or risk perception. The pattern of adoption among large and small farms is generally consistent with the proposition that small farms may lag behind larger farms in the early stages of adoption but soon catch up. The impression from these studies is that the most pervasive explanation of why some farmers do not adopt new varieties and fertiliser while others do is that the expected increase in yield for some farmers is small or nil, while for others it is significant, due to differences (sometimes subtle) in soils, climate, water availability, or other biological factors.
>
> Agricultural technology is more site-specific than we were led to believe by some of the early successes with wheat and rice varieties. The early high-yielding wheat varieties were

adapted to extensive production areas and were widely adopted. It seems that these early successes will not be easily duplicated on such a scale. In the areas not already dominated by new varieties, the factors limiting yields are so disparate and complex as to make it unlikely that any single new variety can repeat the success of the early releases from international breeding programmes.

Government policies to reduce the cost of information, the cost of inputs, or the impact of risk could be expected to influence the decisions of those farmers whose expected yield increases are now marginal. But the experience in the areas of these studies suggests that these policies are not likely to have a large impact in increasing the numbers of farmers who adopt new technologies. Significant advances in adoption will not occur until significant advances are made in technologies that will increase yields in the agro-climatic environment of those farmers not presently adopting. This will require greater attention to the environment for which technologies are being developed than was necessary for the early successes in cereal breeding.[13]

2. IRRI (International Rice Research Institute), Philippines

As with CIMMYT, IRRI's FSR approach came from an observation of the problems of technology adoption and performance. Although it is FSR methodology that is used, IRRI state that their focus is on work with multiple and inter-cropping of rice rather than the broader interests of 'farming systems'. Centred in East and South Asia, where land is limited and yields per area are already high, research is directed at small farmers with the objective of increasing their efficiency of resource use – particularly by intensification of cultivation. In 1974 the Asian Cropping Systems Network (ACSN) was established in an effort to link IRRI's Cropping Systems Programme (CSP) to those national research efforts for which they had been demonstrating support. There now exists a network of 25 locations in seven countries in

South and South-east Asia, where test sites involve farmer cooperators ('Agronomic' cooperators) working with cropping patterns, and 'Economic' cooperators collecting farm records.

CSP researchers use a particular set of procedures, the components of which are (1) environmental description, (2) cropping pattern design, (3) cropping pattern testing, (4) component technology, and (5) pre-production testing (IRRI, 1978). In spite of this approach (which should take into account the socio-economic variables and be adaptive, inasfar as the component technology and pre-production testing procedures analyse and modify the proposed technology, and examine alternatives or additional areas of importance)[14] ex-post-evaluation of rice technology performance has indicated that there exist major differences between yields from farmers' fields and those from the experimental station (IRRI, 1979). The analysis so far suggests that it is not only the recommended varieties and practices that may not be suitable to the small farmer's environment – but also that socio-economic factors are having a direct impact on adoption and performance, e.g. access to credit, supplies etc. The focus of the CSP has previously been one of commodity production and could thus explain IRRI's apparent failure in designing suitable technologies. They have been successful in some areas, e.g. the Kabsaka and Kasatinba programmes in the Philippines (where double rice cropping was introduced), where the agronomic practices appear to have been well adapted to existing farming systems.

The technology that IRRI has attempted to introduce has had mixed results. Their concentration on rice and its production as a commodity has probably been beneficial as it enables the CSP to focus sharply on that specific area. However, that this should be so at the expense of the system for which the technology is being designed has led to certain problems already discussed – something that today IRRI, in conjunction with the ACSN, is trying to remedy. Perhaps one of the most important factors to come out of these experiences is that of establishing institutional linkages within a research organisation. The significance of this lies in the possibilities of widening the programme area, without detracting from each institution's priorities or causing them to enter into areas outside their official mandates. Potentially interinstitutional linkages could have most positive effects on the integration of R and D projects, but judging by the experiences of intra-institutional information networks this potential is not going to be easily operationalised.

3. ICRISAT (International Crops Research Institute for the Semi-Arid Tropics), Hyderabad, India.

ICRISAT is attempting to link two research programmes which are running at present – the Farming Systems Programme (FSP) and the Economics Programme (EP). Essentially the former is concerned with technological development aimed at the improvement of 'land and water management systems . . . to contribute to raising the economic status and quality of life for people in the semi-arid and arid tropics' (Technical Advisory Committee, 1978). The Economics Programme is seeking to identify those factors constraining agricultural growth and to develop both technical and institutional solutions.

Methodologically the two projects differ in that the Farming Systems Programme uses a deductive or 'upstream' approach[15] while the Economics Programme is inductive (or 'downstream') in its orientation. Initially concentration in the FSP was in the investigation of single production components (agro-climatology, hydrology, environmental physics, soil fertility and chemistry, farm power and equipment, land and water management, cropping systems and agronomy and weed science) (Kampen, 1979) with the work being research station based. The EP however – divided into Production Economics and Marketing Economics Programmes – was village-based. The dialogue was poor between the two groups, research being divided along traditional disciplinary lines with little or no linkage of results to influence the direction of technological development. However, more recent events demonstrate that the complementarity of the programmes has been recognised not only as existing but also as being essential. Pressure from funders, who want to see viable ICRISAT technologies spreading on the farms of resource-poor farmers, have been a significant force in promoting more collaboration between social and natural scientists.

Institutionally, ICRISAT provides an example of change through experiences. For a large institution it has demonstrated flexibility – albeit slow to manifest itself – and an ability to respond to the move towards interdisciplinary research and technological development.

Weed problems

Rather than being a synopsis of a case study this part of the paper outlines briefly the attitudes of one of ICRISAT's projects to the weed problems of smallholder farmers in less developed countries. It is useful for us to understand their approach as it has developed from their experiences with, and comprehension of, the farming systems of the small cultivator. It demonstrates how an agency, after identifying a particular technological area, is able to

(a) place this in a system; and

(b) link this to other practices/technologies in that system.

Characteristics that distinguish smallholder farmers in the less developed countries from those in developed countries are considered to be:

1. Tropical weed problems are more complex than those of the temperate zones.
2. The farm sizes are small; in addition, almost all fields are surrounded by uncultivated areas or bunds which are serving as sources of weed infestation.
3. The heterogenous nature of soils makes them difficult to handle.
4. Rainfall is erratic in terms of the total quantities of precipitation and distribution, making tillage difficult and crop growth extremely unpredictable.
5. The technology required, especially for upland crops under rainfed conditions, is complex and, if available, has not yet reached the farmer.
6. The farmer owns limited resources and operates with scarce and expensive capital. The average holding in most cases is less than 3 ha.

7. Most farmers are illiterate and, therefore, much of the available agricultural research is not being understood and adopted by them.

8. Agricultural production is unstable and yields are low.

9. The supply of labour is abundant at the present time.

10. There is lack of understanding by the researcher of the farmer and his production system.

11. Due to the possibilities of intensive cropping the production potential of the small farmer per unit land area is greater than that of larger farmers of the developed world.[16]

It is noted that weeds are only recently beginning to be perceived as a problem by peasant farmers – who tend, if any weeding is to be done at all, to weed late and by hand.

An obvious technological solution to the weed problem for those involved in modern agriculture is the use of tillage techniques and herbicides. However, attempts to introduce these to the small scale producer have been unsuccessful on several counts:

1. Ignorance of the losses caused by weeds and a fatalistic acceptance of weeds – ironically, because of their almost universal presence.

2. The limited cash turnover preventing the adoption of herbicides and other new practices. The income of the farmers is low with little or nothing left to invest in the farm. Most Indian farmers are primarily subsistence farmers marketing little of what they raise.

3. The present day ample supply of labour – both family and hired. The increasing wage rates coupled with the rapid expansion of industry may later result in scarce farm labour, but this is not expected to happen very soon.

4. The farmer usually hesitates to accept risks by adopting new practices. He does not change long existing, well tried, farming practices suddenly. Furthermore, his inability to read and understand the labels of herbicides could prove disastrous.

5. The farmer does not need a "clean" crop and in some instances wants the weeds to become fodder to feed his cattle. Some weeds are deliberately left because they serve as vegetables. Clean culture with total dependence on herbicides may add to the problem of soil erosion on steep and shallow soils.

6. Since the small farmers grow several crops several herbicides may be needed. The problem of growing more than one crop (multiple cropping) in a year becomes more complex when more than one crop is grown simultaneously (intercropping) in one season on the same farm. Obviously there exists an opportunity for mistaken usage of herbicides with serious crop damage. Also very little is known regarding herbicides safe for multicropping and at the same time effective against several weed species.

7. The lack of trained personnel and weed scientists who have undergone training in Weed Science. At present, agricultural universities, research centres, governmental extension departments, and pesticide industries do not have sufficient technical know-how in the field of herbicide science. Only very few universities have included weed science in their teaching curriculum.[17]

It is considered by ICRISAT that despite these problems weed control is necessary if agricultural production is to be increased. Herbicides are still proposed as the answer – but with the proviso that research should take into consideration the complete environment of the farmer. It is suggested that such an approach would incorporate the development of simple, short-term herbicides, minimal dose and 'hand' application, simplified equipment, adaption of existing systems and low-cost technology – specifically that suited to local soils, climates and cropping patterns:

> Therefore, the research approaches should be to combine the principles of the existing cropping systems with available new technology to develop effective systems of weed management. We should depart from the conventional 'chemical only' approach to an integrated management approach.[18]

This approach tries to 'tip the balance' in favour of the crop in terms of economic value. This can be done by using three means in conjunction with one another:

(i) systems of cropping methods;
(ii) cultural practice techniques, and
(iii) the *supplemental* use of herbicides (Binswanger, 1977).

The evolution of such an approach should take place on the farm with constant reference to the idea that weed control is part of a system – not an isolated entity in itself.

An example of a system that has hopefully used the above criteria is the broad ridge and furrow system (Krantz, 1976), developed in an attempt to manage (a) the weedy field bunds, and (b) the weeds left after harvest:

1. Broad (150 cm) ridges and furrows are established on a graded contour with an average slope of 0.6 to 0.8% in black soils and 0.4 to 0.6% in red soils. These ridges and furrows lead to a natural drainage way which should be grassed to prevent gulley erosion. Thus once these land features are established they can become permanent just like graded contour terraces and no additional surveying or 'laying out' is needed. All future cultural operations follow and maintain the broad ridges and furrows. The ploughing, ridging and planting is all being done with an animal-drawn, two-wheeled tool carrier. In all operations the wheels and bullocks always follow the furrows; thus there is no soil compaction in the broad ridge crop production area. Since the broad ridge and furrow system can be permanent land features, the land area of each farmer can be designated by the number of broad ridges he owns and can be marked with a granite block or stone. Thus, the need for field bunds is eliminated and the system would enable each farmer to gain up to 10% of land area previously occupied by bunds, and would reduce the weed seed problem.
2. By maintaining a continuum of crops from the onset of the monsoon through the rainy period and as far into the post-monsoon period as possible by various means of double cropping, weed growth can be greatly reduced by effective crop competition after early weed control.[19]

In summer then, ICRISAT is attempting to utilise Farming Systems Research procedures in its weed research. It is considered that this interdisciplinary, farm level approach is the only one that is going to provide a weed technology appropriate to and therefore adoptable by the small peasant farmer. Economic issues, from the smallholder farmer's point of view, are seen as the starting point of the analysis. While committed to an interactive, interdisciplinary approach, how far ICRISAT has been able to put this sensible approach into practice and develop viable farming systems remains to be seen.

4. ICARDA (International Centre for Agricultural Research in the Dry Areas), Syria.

In 1977 ICARDA isolated three areas that were felt to be of importance to their future programmes:

(1) There was little known about the dynamics of existing farming systems.

(2) Knowledge of productivity was imperfectly understood.

(3) There was a lack of confidence among some of the research community in the conventional approaches to research.[20]

It was therefore felt that there was an obvious need to rethink and clarify the research objectives of ICARDA. Five programmes were already in progress looking at rainfed food and forage crops and at livestock production. The focus was on Cereals, Grain Legumes, Forage, Farming Systems and Training and Communication. However, as Gibbon points out:

> The Centre itself was established in an area (N.W. Syria) where there was no reliable recording system or existing farming systems and ICARDA scientists were not initially linked to the farming community in any way. It was felt that these links and the setting up of a continuing study were vital to the operation of feasible alternatives.[21]

By 1978 an interdisciplinary FSR group had been established, and from the primary task of examining existing farming systems evolved several projects:

1. A study of agriculture within the Syrian economy.
2. Studies of existing farming systems.
3. Climate/soil water and nutrient/crop inter-relationships and water and nutrient management.
4. Cropping systems and crop/livestock systems.
5. Farmers' field trials.

The technical recommendations arising from these projects were interesting by their absence. The previously recognised polarisation of crop and livestock systems and of irrigated and rainfed systems was not apparent, and the accusations of periods of unproductive fallowing by farmers were found to be void – the fallow, in fact, being an important factor in the maintenance of soil stability. Management practices that had previously been criticised by researchers were realised to relate directly to the options open to farmers. What was gained by this particular programme was the generation of an approach to research that helped those involved in its processes to put their work in 'the context of real farming situations' (Gibbon, 1981). As is stated in an earlier paper:

> In developing a research programme with the main objective of improving the productivity and stability of legume crops, it is essential that an approach involving the consideration of the wide range of variables that affect farmers' capacities to influence production and income be adopted in preference to a fragmented approach considering only the legume crops themselves. Through studies of the whole sphere of farm and household activities, an understanding of the range of choices open to farmers and the formation of farming strategies as related to the development of the family may be achieved. The family, however, should not be the only unit of analysis; comparisons must be made between family groups, between agro-ecological zones and between major production areas to fully understand the linkages between local and natural structures and the farmers' extent of control over resources. Decision-making and problem definition at the village level has appreciable consequences for policymaking at the governmental level. For this reason, an understanding of the relations between production in different areas, which can be achieved through a thorough comprehension of the distribution of resources and income and the utilisation of labour, should be considered to be an important part of such an approach.[22]

Again the lessons have been those of necessary specialisation, but with strong institutional linkages, from those with the farmer, through to those crossing disciplinary borders to links with other agencies and institutions.

5. IITA (International Institute of Tropical Agriculture), Nigeria

IITA's 1978 Annual Report states that their primary objective is in:

> Developing methods of crop management and land use suited to the humid and sub-humid tropics which will enable more efficient and sustained production of food crops to be both technically and economically feasible in these zones.[23]

Their programme, unlike those already discussed, is not one of commodity, but of management, focus, and because of this primary concern the content of IITA's research has necessarily been diverse. Most of their work has been carried out at the research station in the Savanna zone of West Africa. Consequently many of the results have not been tested under on-farm conditions. They stated in 1979 that they considered their Farming Systems Programme had the following five components:

(1) Regional analysis involving analysis of farming systems of the region to identify potentials and constraints on production.

(2) Cropping systems involving development of improved cropping practices and alternative systems of crop management.

(3) Land management involving development of improved methods for land clearing and soil management.

(4) Energy management involving development of implements and methods to relieve energy constraints to crop production and processing.

(5) Technology evaluation involving developing, testing, and evaluating improved practices and systems.

Thus there is obviously an awareness of the need for farm-level studies and testing.

The increased emphasis on the FSR approach has led to sharply focused technology programmes with the specific aim of benefiting the small farmer. The Ford Foundation, in recognising this attempt, has recently granted $310,000 to IITA to enable them to introduce agricultural scientists from Nigeria, the Ivory Coast and Cameroon to FSR (Ford Foundation, 1982). The greatest change in IITA over the past few years has thus been the realisation (and consequent structural reorganisation) of the need for an interdisciplinary and farm level focused approach.

6. CIAT (Centro Internacional Agricultura Tropicale), Colombia

CIAT has a fairly narrow commodity oriented programme looking at increased productivity of cassava, beans and improved and new pasture for poor areas in tropical regions of Latin America. Secondary projects are being carried out in collaboration with CIMMYT (maize) and IRRI (rice). Other linkages have been made with national programmes to provide local sites and testing for their small-farm focus technology.

FSR principles have been carefully adhered to, with emphasis being placed on (1) the relevance of technology to the client group (small farmers), (2) on-farm surveys and testing, (3) studies of the economics of the technology for different size farms, and (4) ex-post analysis of effects, adoption and distribution. CIAT is using its findings to influence future research directions and programme planning.

The previous examples have illustrated the attempts and problems faced by international agricultural research institutes to introduce interdisciplinary analysis into their own work and to develop new methods and approaches to research on small farmer problems.

7. ICA (Colombian Agricultural Institute)

The Caqueza Project was the initial attempt by ICA to work with FSR procedures in a rural development context. The lessons learned there such as the difference in problem identification by farmer, planner and politician; the need to develop accessible technology and the importance of locating the technology in the production situation, have to a great extent shaped the institutional thinking and methodological approaches to research in Colombia. The overall programme objectives were to:

(1) develop and prove a strategy for the transference of technical, economic and social knowledge to small farmers that would promote their active participation in matters such as the use of credit and purchased inputs, the sale of their products, and the betterment of their social conditions;

(2) use this strategy to bring about higher crop and animal yield, improved economic returns, and better family living in the project area;

(3) establish a system whereby the farmers of the project area assumed increasing responsibility for the execution and expansion of this strategy using their own initiatives; and

(4) measure the changes in the community, and in its incomes that resulted from the project.[24]

The International Development Research Centre (IDRC) of Canada assisted in this essentially interdisciplinary programme. The weight given to the 'bottom-up' strategy became increasingly important:

> As the project developed, the first two objectives became closely interwoven with the strategy being developed, tested and propagated on farmers' lands. The project staff never had experimental lands at their disposal and all research had to be conducted at the farm level. To bring this about a new approach to production research was developed. This approach seems likely to have important long-term effects on Colombian agricultural policy as it has been adopted at the national level.[25]

Conclusions as to the success of the project are difficult to draw, as the programme was a demonstration of both research *and* development – the results of which it was hoped would have an impact on similar client groups elsewhere in Colombia. Therefore any analysis has to be on two levels – success in terms of the Caqueza area, and success at the national level.

In terms of the project area the client group of small low-income farmers was successfully identified, but institutional problems (due to the fact that ICA was primarily a *research* institution) led initially to much of the technology that was developed being inappropriate for them. Fortunately, as the project was one incorporating constant revaluation and reorientation, this problem was overcome. The dissemination of the technology developed was achieved through working with small groups rather than individual farmers, and results are reflected in the rates of adoption of the various technologies offered. It was found that for potatoes and vegetables acceptance of varieties and management procedures was reasonable, provided the markets for this produce existed and that prices were right. Producers however:

> . . . in the absence of credit or labour constraints, would not follow proven recommendations on a crop which could be grown without incurring cash costs, even when net income increases of over 20% had been demonstrated.[26]

Attempts to solve this apparent problem led to the discovery of the real cost of credit to the small farmer and an explanation of their behaviour in terms of risk avoidance. It is difficult to judge the benefits of the project in terms of improving the living standards for the large number of farming families in the region:

> The experience of the Caqueza project is that much of the technical effort invested into agricultural production research may not necessarily improve the well-being of the small farmer, unless the technical innovations are accompanied, at least in the first instance, by a comparable expenditure on activities to investigate the farmers' limitations to the incorporation of research results into their production system.[27]

The institutional experiences of the project have thus been of great value in the reorientation of agricultural research and in the demonstration of the desirability of taking and incorporating FSR in rural development projects. As Zandstra states:

> After five years of activity the project continues its learning process. Some of its earlier experiences have, however, been translated into field programmes. The most important lesson learned from the project is that there is often a considerable difference between what the farmer really wants and what the politician and planner feel that he needs. The early emphasis on developing and transferring a new high-production technology has been replaced in the research programme by an effort that seeks to make improved technology available to the

farmer. The farmers have constantly stressed that what they require are packages of technology that they can afford to utilise, given the price and credit framework available to them, that are associated with a marketing system that is sufficiently responsive to demand changes so as to encourage them to use new technology to achieve higher production. They have rejected, and will probably continue to reject, technology offered them in isolation of credit and marketing links.[28]

8. G.B. Pant Agricultural and Technology University at Pantnagar, India

(a) Triticale

The case in question here comes under the All India Coordinated Wheat Improvement Project and their programme for developing triticale for both hill and plain areas. As will be illustrated by the following description, researchers became uncomfortably aware, during the programme, of the inappropriate way in which their research and analysis of the problem had taken place. They had a preconceived idea of the technological solution, but during attempts to develop technology to encourage farmers to adopt this it became obvious that this 'top-down' approach was not a cost-effective use of time and resources. It is from such an experience that agencies can benefit by accepting that diagnosis of farmers' problems can only come about through going into the field to work with, and understand the conditions of, clients. Another lesson is that, having been able to acknowledge a mistake, an agency/organisation needs the professional humility and flexibility to re-assess and change the direction of their programme. This is a process that took place at a field workshop reviewing triticale data and the results of on-farm trials.

The case: it is with the hill areas and the very poor smallholder farmers that our interest lies, mainly as it was here that the focus of the project lay:

The triticale team identified as their client group all those farmers in Uttar Pradesh who might grow triticale during the winter season in the Himalayan foothills. Most of the land holdings are small and highly fragmented, with 49% less than 1 ha. according to the 1970-71 census. Sixty per cent of all cultivators are women. In a detailed study by Singh and Rahim of 10 villages in Bhikasen Block in Almora district, the average size of a land holding was found to be 0.47 ha; this is probably representative of much of the region.

Despite differences in attitudes towards risk and access to credit, fertilisers, etc. which this type of skewed distribution of land generally implies, it was thought that new triticale varieties with consistently higher grain yields than alternative crops under non-irrigated, low fertility conditions would benefit all farmers. The added stability of yield was thought to bring an additional advantage to the small farmers who make up the majority of hill cultivators, as they could least afford to take high risk.[29]

In some trials mounted on experimental stations of the Himalayan foothills triticale gave yields of at least 20 per cent more grain than the traditional varieties of wheat and barley grown (Srivastava, 1973). Subsequent to these results a demonstration programme was extended and a mini-kit scheme implemented. The results of these promotion schemes were poor.

To investigate whether triticale was spreading, and, if not, why not, a programme of on-farm trials, surveys and communication methods was started by the scientists. Three types of on-farm trial took place in 1978-79 (1) a varietal trial (different triticales versus a traditional winter crop); (2) a time of planting trial (an attempt to understand problems of sterility at high altitudes); and (3) a fertiliser trial (response of triticale to fertiliser under farmer conditions). These were accompanied by three farm and village level surveys, the first looked at the conditions of those cooperating with the trials, the second at the experiences of farmers who have grown triticale at some point. The third was a triticale potential survey attempting to determine the various production characteristics and problems of hill farmers.

From the trials in became apparent that traditional varieties and other crops available to farmers often had a yield potential as high as those of the new crop, similarly they were far more suited to the agro-climatological conditions of the highland. Fertiliser response of local crops appeared good, but in terms of agricultural development it is the supply and cost of this factor that is of importance:

> To understand these issues adequately, farmers would have to be asked about current cropping and fertiliser practices, relative costs involved, expected yields, etc. and whether they have access to information about, and supplies of, fertiliser. In addition, a programme of accurate crop cuttings of local varieties in the hills would also be needed, together with a systematic collection and testing programme for local cultivators to determine their actual yield potential.[29]

As this, and the following example illustrates, a critical institutional lesson is that a minimum of research funds in major plant breeding programmes is needed in order to allow them to conduct their own on-farm and village level research programmes. Without this component, which facilitates an on-going dialogue with former clients, a breeding programme may misspecify technical priorities.

(b) *Maize*

In the All India Coordinated Maize Improvement Project there has been a major effort to improve varieties in North India. A major focus has been on the problems of maize growers in Western and mid-Western Uttar Pradesh, who cultivate 22 per cent of the 5.7 million ha. under maize in India.

Maize is an important crop to the area, often double cropped with potatoes or wheat, or intercropped with pulses, as a risk-averting strategy. It is grown mainly as a subsistence crop. Farm size is generally low – 90 per cent of holdings under 3 ha. (1971 census), and sharecropping is widespread.

Demonstration plots had been set up for many years to show small farmers the possible results from a package that included hybrid maize varieties, composite varieties, fertiliser inputs and pesticides. Results achieved on these plots were of yields of 3 tonnes per ha. However, the overall yield for the crop in Uttar Pradesh remained at 1 tonne per ha. and it was reported that adoption of improved varieties was poor.

It was decided by the research station scientists that on-farm research was necessary. Projects were set up in two districts where cropping patterns were different. The Bulandshar area had very few farmers growing hybrid maize, whereas in Moradabad a number of producers had been cultivating it for several years. Varietal trials, fertiliser trials, verification trials and 23 production plots were set up. A one day exploratory survey, a small random sample agronomy survey and a marketing and utilisation survey were planned. Communications between scientists were identified as an important issue, and three methods were utilised to assist in the flow of information. Firstly those involved in on-station research had the responsibility of designing and analysing results of the on-farm trials and surveys. Secondly, farm sites were to be visited by local national and international scientists. Finally all results and comments were to be published before the next season.

Extensive flooding during the monsoon of the first year of the on-farm research programme demonstrated that local varieties could give better yields under such conditions. One of the technical problems was identified as one of non-synchronisation of tassels and silk emergence under stress conditions. As a result of this on-farm research finding the screening standards were changed in on-station breeding. It was also discovered that leaf-spot disease was a greater problem for HYV's than was previously thought, as was late wilt and sugar-cane downy mildew. On-farm trials confirmed an important policy decision that short term maturation of maize was far more important than the longer maturing hybrids that could give higher yields. Additionally it was found that the environment of the major research station at Pantnagar was very, very different from the conditions faced by most farmers.

Weeds were discussed as a possible constraint on production but further investigations illustrated the labour cost for weeding for smallholder farmers was very low, therefore labour intensive methods of weeding were a far more appropriate avenue for research than any capital demanding methods such as herbicides and mechanical weeders. However, some visiting experts were suggesting research on capital intensive techniques as large farmers wanted to reduce their labour bill.

Both the triticale and maize on farm research programmes, conducted in conjunction with CIMMYT, revealed that the adoption of new varieties and agronomy recommendations had been low because farmers' problems had been inadequately diagnosed. This had led to research resources being misallocated.

The target groups in this case benefited little from the technology they had been subject to – and the questions remain those of (1) was the technology identified either necessary or the correct one? and (2) what are the true constraints on production in this area? As Biggs states, for the triticale research and extension programme:

> However, within an integrated natural and human resource planning framework for the hills and plains there may still be justification for the development of suitable crops, such as triticale for the land which remains under cereal cultivation.
>
> Perhaps the most significant and important lesson to be learned from this case study is the overriding importance of the strong and necessary linkage between the farmer/client group and the scientists. It is not only a matter of recognising that information about the agro-economic conditions of farmers is important, but one of ensuring that the information is used to influence research priorities. In this case, therefore, it is significant that the generation of new information through trials, surveys, and a field workshop, and the discussions on the implications of the analysis all took place within the local triticale programme.[29]

Experience here again demonstrates that perhaps the actual beneficiary group of the R&D and extension programmes were not, in fact, the declared client groups (the poor hill farmer and the smallholder maize producer) but the research teams themselves. Although subject to no outside scrutiny or reports, the results of their current decision to introduce on-farm research have made it self-evident that a change in the direction and focus of their work is necessary.

Perhaps this change will also incorporate a revaluation of the type of crop that FSR has concentrated on in the past. There has been a narrow focus by the major research establishments on crops such as maize, rice and wheat. What is becoming increasingly apparent in situations of land shortage and population stress is that there is a need for research on the traditional crops, especially those grown for subsistence or survival. Apart from a very few small agencies there is a deficit of information that could perhaps lead to more efficient production and storage of such crops as cassava, yams and sweet potato.

Historically in West Africa the emphasis has been on developing export crops, e.g. oil palm, cocoa, rubber, cotton, groundnuts, maize, etc. However, land previously able to support traditional cultivators has become increasingly scarce as the population has grown and more has been taken for production of cash crops. Urbanisation meant that certain food crops fell into this category, and government money was allocated to research aimed at increasing the supply of these products. However, the benefits of such a policy have generally accrued to the middle and large farmers, sometimes serving only to exacerbate the problems of the smallholder and subsistence farmer. It is this group that are in need of help now, and not only in terms of generating a cash income for their families, but at the very bottom of the scale, in terms of staying alive. Other technological areas are of donors' importance here, but in relation to crop research it is essential that work is put into the crops that are able to maintain the very existence of the poor.

B. GOVERNMENT AGRICULTURAL DEPARTMENTS

1. Indonesia

Following the failure of an earlier plan, the government in 1963 launched a new attack on the possibilities of raising the rice yields of small farmers in Java.

This case is one where, initially, a useful and correct approach to the diffusion and flow of information was taken, and serves to demonstrate how political conditions can influence a programme in spite of an awareness by planners and participants of the detrimental effects of this interference.

The technological package was one of fertilisers, new varieties and weed control – but the method of transmitting the information was different. It was one based on the assumption that without any structural changes, i.e. land reform, rice output could be increased through a change in the extension work:

> The architects of the new method took a calculated decision to trade-off expertise against enthusiasm and chose 12 students of the Bogor Agricultural College for their pilot projects. The student extension workers were obliged to teach farmers in groups of 8 to 12, to live continuously in the village, and to work alongside the farmers in the field. They had to prepare something less than one hectare of land as a demonstration plot.
>
> The results of the pilot project were encouraging. The scheme was expanded. During 1964 and 1965 student extension workers took it upon themselves to find out about supplies of fertilisers and to harass and threaten cooperative and other officials they felt responsible for maldistribution of the inputs.
>
> The new programme thus had a new approach to farmers: an ideology of new farming methods; credit to purchase a 'package' of modern inputs; and intensive, grassroots guidance. The principle of two-way traffic was to be upheld at all times.[30]

Initially results were encouraging and the programme expanded. However, several new factors came into play – although between 1965-66 1,200 students were placed, the area they had to cover was increased, it was decided that no more credit would be available to previous bad debtors and the political climate was such that students would have been risking themselves in a continuation of the 'harassing of malfunctionaries' (Palmer, 1975). There were problems too with supply of inputs.

By 1968 the government recognised the scheme was no longer achieving its goals and turned towards the multinationals in an attempt to realise their objectives. The subsequent package was very rigid, and was abandoned in May 1970 when massive social unrest appeared imminent. The following packages were technologically similar to those having gone before, but with an emphasis placed on easier access to inputs and greater flexibility allowing for the individual's production decisions. It was a system incorporating the use of coupon books – thus simplifying problems of access. Credit facilities were improved (with the use of mobile banks) and fertiliser kiosks were set up.

The success of these government projects has, in terms of the programme objectives, been minimal. The smallholder poor farmer remains in much the same position as he was prior to the 1960s, and his rice output has increased by a negligible amount (Palmer, 1975). The earlier lessons of how an institution can interact with their defined beneficiary group, and where a certain amount of positive response was demonstrated by the farmers, should not be forgotten. They served to illustrate the importance of working with the client group at farm level, being part of, and understanding, their decision-making environment.

2. The Gambia

Launched in 1971, the Mixed Vegetable Scheme in The Gambia was a cooperative effort by the Ministry of Agriculture and Natural Resources, the Gambia Cooperative Union and Freedom from Hunger.

The target was twofold. By introducing improved varieties of onion it was hoped (1) to reduce the reliance on imports, and (2) 'to create an income-generating role for women and to open up an opportunity for women to form cooperatives and become part of the Gambia Cooperative Union' (Morss, 1976). Although most of the women participating in this programme were not those from poverty groups alone, it is felt that this programme is of value in terms of the lessons it demonstrates, i.e. defining the group at which the programme is focused; how to work with the beneficiary group and how to utilise existing systems – such as the women's familiarity with vegetable growing and cooperative labour.

After initial successful village testing on a suitable plot –

chosen by the instigator of the idea, a British expatriate – the programme went ahead:

> The technological package was relatively simple, and involved clearing the land, constructing the beds (24 by 3 feet), fertilising and planting the onions in rows, thinning, and harvesting. The total of about 12 man-days was required to cultivate ten beds, which represented a small portion of the time women normally spent on farm labour.[31]

The whole scheme was based on the principle of cooperation with the extension officer working with and through the village chief, and the women forming pre-cooperative groups. Vegetables had traditionally been grown on village communal land, and although the same area was used the land was now divided into private plots.

Previously marketing of vegetables had been problematic due to the infrequency and high cost of transport. Under this programme the onions were collected by an agent of the Gambia Cooperative Union who paid the women a fixed price. Problems arose over the discrepancy in the Banjul market price and the monies received by the women. However, participation has not decreased as the profitability of the venture, even with fixed pricing, is recognised.

What effects has the project had? Unfortunately, due to bad planning the government continued to import onions, and flooded the market so in terms of the first programme objective the results were not good. However, the producers were protected by their price fixing agreement and therefore did not suffer. This is an interesting example of how responsibility for the technology was accepted and given practical and legislative expression. The financial benefits accruing from the introduction of this simple technology were those of increasing an average farm income by 7 per cent. Several reasons for the success of the project can be drawn out:

> The requirements of the project were minimal in terms of labour, land (provided by the village chief), and other family resource commitment.
>
> In addition to having direct income benefits, the project has begun to institutionalise the role of women as innovators. Gradually, the pre-cooperatives are building management and

leadership skills and have the potential for conducting other income-producing schemes. Moreover, the improved techniques, particularly in the case of Busumbala, have been applied to growing other vegetables.[32]

C. NON-GOVERNMENT ORGANISATIONS

Note here that in many cases local NGOs have worked in cooperation with agencies that have been categorised under other headings.

1. FUNDAEC (Foundation for Application and Teaching of Science), Colombia

The work of this group in Cali serves as a case to illustrate the importance of two issues. Firstly it stresses the point that it is important for such small agencies to work with farmers on the improvement of crops grown for home consumption and subsistence, and secondly it reaffirms the constant need that this paper outlines for working with and identifying correctly the beneficiary group for which a given technology is intended.

Sugar-cane is the main crop grown in the Cali area. Grown commercially as a monocrop, it demands high technology inputs such as fertilisers, herbicides, irrigation, etc. Consequently differentiation within the farming community, in terms of ownership of land, has increased. Many local people have become wage labourers on the plantations. FUNDAEC's interest in the area has been twofold. Primarily they wish to assist the local people in more efficient production from the small plots they still hold – thus (a) reducing their dependence on wage incomes, and (b) increasing their nutritional status.

The other objective of the programme is to reduce the soil erosion that the intensive cropping of sugar-cane is causing. Thus immediately two user groups have been identified – the plantation farmer, and the small-scale subsistence producer. The interest in terms of this paper has been with the latter group:

> FUNDAEC's staff of agronomists and rural engineers is conducting field trials with different crop varieties, fertilisers, planting schedules and the like. Most of this work is done directly with local farmers, who learn how to maintain their own small integrated systems. The farmers are trying intercropping, fish ponds, fruit crops and other ideas. The response from most of them has been very good.[33]

The technologies being developed are those specifically in response to the conditions and constraints found in the Cali area. Intercropping of corn and plantain has been one management technique introduced. The growth of these crops together is beneficial as (a) it is efficient in terms of space sharing; and (b) it reduces bird damage.

Leguminous plants have also been introduced – to be planted after harvesting of these crops – to assist in nitrogen fixation, and to provide a good source of secondary protein for the farming families. Other simple techniques have been presented in the areas of land and management techniques, such as integrating small-scale animal husbandry with a fertiliser programme and introducing fish and chickens as a new input to the food chain.

The projects appear to be having a real impact on the lives of the small producer (Corven, 1982) and their effectiveness can be attributed to the farm level identification of problems by the agency, the direct involvement of farmers in the research work and the development from that of technological assistance wholly appropriate to the given conditions. The technology itself often takes the form of advice rather than hardware.

2. Lirhembe Multi-Service Cooperative, Kenya

Set up in 1972 the project had two objectives:

(1) To increase agricultural production of maize, vegetables, passion fruit and livestock, and

(2) To improve social services available.

This example is one of where a service centre was set up, serving to act as a point from which services were distributed to, and procured by, the identified client group. How and why this particular centre functioned successfully is discussed below.

The client group were the very poor living in the 1,000 acres that constitute this area. The average farm size is 2.5 acres, and in spite of previous extension attempts, farmers continue to grow local varieties of maize and vegetables and raise livestock for home consumption. The literacy rate in the area is roughly 60 per cent, and many of the men have left the area to gain off-farm employment in the cities.

The first stage of the project involved the construction of a social centre – which was used for both leisure and educational activities – including agricultural extension operations. Demonstration plots, a cow dip, milk cooler and maize storage facilities were all located close to the centre.

The introduction of grade cattle (in calf) and the use of the cow dip (farmers cease to become members of the co-op if they miss the required sessions) had by 1974 led to the death rate of cattle falling to 6 per cent. In the neighbouring district the mortality rate of cattle is 60 per cent. There has been an interesting shift in the structure of household decision-making since the profitability of the new package has been manifest – women are beginning to make decisions regarding credit (for purchase of cattle, use of dip, and preparation of pasture) whereas previously men had dealt with all major production credit issues.

During the first year of new-variety maize cultivation, farmers were given seed and fertiliser free of charge. In the second year the package of improved seeds, fertiliser and insecticides was offered interest free with the proviso that all produce was sold to the cooperative. Repayment rate in the first year was 95 per cent, but during bad weather in the following year this figure fell to 30 per cent. In attempts to reduce this high rate of default and find methods to ensure repayment, the leaders of the project have initiated *barazas* (local meetings) at which the problems can be discussed.

Passion fruit and vegetable technologies were introduced in a similar manner, with the objective of market production. However, outlets have not been arranged – which could explain the low level of adoption.

The project has been successful in its aim of increasing family income and also in improving the level and quality of social services, albeit at the possible expense of other areas. It is estimated that at least 70 per cent of families have adopted, at least partially, the technology offered. Farmers outside the area have also joined the cooperative to take advantage of the maize package offered.

The reasons for this success are that the project was conceived, planned and implemented locally, although it was granted $140,000 by the Dutch charity NOVIB who proved flexible in their approach to the way in which the money was used – leaving decisions to the project coordinators and participants. Local leadership was strong and the farm level involvement of those planning the exercise high. Physically the area was very small and the people already formed a cohesive community suitable both economically and socially to the organisational development of cooperative exercises. It would appear that the social differentiation between farmers in this area was not large or politically divisive enough to render the cooperative an institution for serving the interests only of the richer groups. Because of these factors, communications at and between all levels were good. The social centre provided a focal point for the programme, and its use for a diverse number of activities appears to have afforded a physical manifestation to the ideal of the cooperative venture.

3. ASAR/ARADO Potato Production and Seed Improvement Project, Bolivia

The Association of Artisan and Rural Service (ASAR) is a technical assistance agency aimed at helping peasant-managed base institutions. ARADO (Acción Rural Agrícola de Desarrollo Organización) are rural peasant based units.

Institutionally the interaction of the two groups should be noted, especially with regard to the recognition of how one group was more suited to a particular role than the other. The mechanics of this process reflect directly the ability (and demonstrate how) an agency can diagnose problems, work and maintain contact with, their identified beneficiary group.

Initiated in 1968, this programme was designed to meet the need for increasing potato yields. Potatoes are both the major subsistence and cash crop of the Cochabamba region of Bolivia – and as yields were consistently low, insect and fungus attacks frequent and credit arrangements unfavourable to the subsistence farmer, it was considered to be the primary target for any improved technology. Although farms in the area average 4.4 acres, soil is very poor, the altitude is high and farming is characteristically subsistence in nature.

The potato production and seed improvement scheme was introduced in 1968, utilising monies granted from the German Catholic Bishop Institute, Misereor and later by Oxfam. It involved an interest-free package of improved seed, seed selections, planting density, fumigation, fertilisation and weeding – all recommendations of the Ministry of Agriculture's research station. Default of loan repayment was high during the first year. The problem was discussed with peasant leaders and members of the organisations and a list of recommendations drawn up. The solutions offered were directed at the use of existing local structures. The credit system was to be similar to the traditional *compania* system, and the responsibility for supply and recovery of credit transferred from the technical organisation (ASAR) to the peasant based ARADO. Both measures have had positive results, and delinquent payments are no longer a problem.

Technical assistance to the subsistence farmer took the form of farmer/technician discussions as to the type of seed and the plot of land to be used in production; the delivery of seed and other inputs to the farm; technical supervision of planting, disease and insect control, harvesting and grading and collection of that part of the crop that the farmer wishes to market.

Since the improvement of dialogue between the farmer and the ASAR administrators through the mediation of ARADO, crop yields have increased substantially (approximately 55 per cent), a fact that is reflected in the increased income of farmers. There are still problems in the application of insecticides and pesticides in terms of both timing and cost of inputs. Seed potatoes tend to be scarce and have proved a major limitation on the spread of the technology.

The programme's success can be directly linked to its flexibility and responsiveness to its client group. At no point were the technical problems of farmers divorced from credit and other problems. It has utilised traditional social and economic systems and maintained an approach of strengthening local grass roots organisations and participation.

D. INTERNATIONAL DEVELOPMENT AGENCIES

Most cases of international agency involvement have been included under other case study agency types; for example cases covering the work of the International Agricultural Research Institutes.

IBRD Agricultural Development Project (International Bank for Rural Development), The Gambia

This case is of particular interest as it is one in which an institutionally determined technology has been directed at a specific group. The peripheral effect of this has been to disadvantage another group in that society, located in the MacCarthy Island Division (MID) of The Gambia:

> The project was identified by the World Bank's Permanent Mission in West Africa in February 1969, and the FAO/IBRD Cooperative Programme completed the preparation of the project in July 1971. During its visit in February 1972, the IDA appraisal mission made several modifications at the suggestion of the Gambian government; its report was submitted in June 1972 for approval by IBRD. Two major efforts set the stage for the Bank's project. The first was the Commonwealth Development Corporation's Mechanised Rice Production Project in the 1950s, which demonstrated the potential for growing irrigated rice and also the problems of designing an effort which was beyond the technical and maintenance capabilities of the Gambians. The second was the Chinese Irrigated Rice Produc-

tion Project which offered a well-tested technological package and which had laid the groundwork for farmer acceptance of that package. Seeing some basic shortcomings in the Chinese approach, the Gambian government incorporated ideas in the IBRD's design to increase the possibility of the project becoming financially viable and self-sustaining. The most important of these ideas were the training and use of Gambians in the project and the provision of credit at unsubsidised rates for all inputs.

Another important feature of the project's design is that it provides for the experimentation necessary for longer-term planning. Starting with one "proven" idea, the project has the mandate and resources to experiment and develop a more comprehensive development programme.[34]

MID stretches for 100 miles along the banks of the Gambia River and has fertile soil that produces 30 per cent of the country's groundnuts (grown under shifting cultivation) and 25 per cent of the rice (grown on the river flats and swamps). Average farm size is 4 acres, rural incomes are low and the literacy rate is only about 10 per cent.

The technological package offered differed from the Chinese model in that it used hand rather than mechanised labour for cultivation, and pumps were provided on credit. Basic agricultural practices of nursery planting, transplanting and tripartite fertilisation and weeding remained the same. Harvesting was carried out by hand and threshing done by small pedal-operated machines. With this package it was hoped to bring 3,000 acres of land under irrigation for rice production over a period of three years.

Production sites were located within village boundaries, and a cooperative formed responsible for its development. Mechanisation was used in the initial ploughing of the land, by government tractors. Farmers were at first reticent to join the programme as they knew that on the Chinese scheme the primary inputs and the pumps were provided free of charge. However, this was overcome when it was realised that the Chinese were not going to operate in this particular area, and that loans would be easily repayable out of the extra income generated. To date loan repayments have been 100 per cent.

Extension facilities took the form of previously unemployed young Gambians, who had been intensively trained, and who provided daily assistance to the cooperatives.

Benefits to the local participants have included an increase in income (despite the higher costs of production) and the utilisation of previously underemployed labour. However, the socio-economic effects of the project on the relationship between men and women have tended to have a negative impact on the latter group. Traditionally upland and swamp rice were cultivated by women as both a food and cash crop. This programme introduced rice cultivation to men, who have since been exercising their ownership of irrigated land and:

> institutionalising an inheritance system that will keep it under male control. It is only with considerable difficulty that a few women acquire use-rights to a plot in the dry season. Women are lent the usufruct over some irrigated rice plots in the rainy season where they grow traditional varieties of rice because the land is subject to tidal flooding and is therefore unsuitable for irrigated rice cultivation. The men, quite simply, do not require the land in the rains.[35]

Women have thus become disadvantaged in that they are no longer able to partake of an income-generating activity apart from the rather irregular, poorly-paid wage labour on the men's plots.

The increase in income that the men have experienced appears to be having little effect on their wives or families. Those women who do display signs of improved living standards have achieved this through their own hard work.

Dey also notes that the rather disappointing low levels of cropping rates can be related to the planners' ignorance of this issue of traditional division of labour. Women in the past have been relied on for their cultivation skills. She suggests that had the planners included women in their programme, 'it is probable that double cropping of irrigated rice would have been achieved on the women's fields at least' (Dey, 1980) and could have utilised their already established reciprocal labour groups.

The programme has highlighted two factors important to the success of projects dominated by international agencies. Firstly, it is important that nationals who effectively represent different beneficiary client groups are included in decision making and implementation, and secondly, the exercise must be flexible enough to adopt a 'proven' technology to local, physical, social and economic conditions.

E. INFORMAL RESEARCH AND DEVELOPMENT

In this category we include technology which has been generated and adapted by the 'informal' R and D activities of local farmers, artisans, etc, themselves. In this case the 'clients' are often the same people as the individuals doing the R and D.[36] The technology diffuses amongst other producers who are similar. Two cases of technology adopted by, and benefiting, specific client groups are:

(i) Rubber-cassava grafting in Java where Ker (1979) reports that farmers have been grafting ceava rubber plants on to cultivated cassava, giving rise to a canopy shade that increases cassava yield up to 100 kg per root, and

(ii) farmers involved in Plan Puebla (see the earlier case study).

Although both examples illustrate 'informal' development of technology, the cases differ in that the former involved a completely indigenous innovation – whereas the latter was a development out of an exogenously introduced package.

1. Bangladeshi farmers

The point made by this study is that farmers themselves will carry out research and innovate according to the results they obtain. Unfortunately most agencies do not recognise the strengths and complementary nature of informal R and D and do not have institutional arrangements to include it in their research processes and on-farm trials:

As regards the specific effects of informal R and D on recent varieties of rice and wheat from experimental stations, we find that new developments have come about because of the informal system. For example, after 'dwarf' IR8 rice was introduced in Bangladesh, farmers were found to be actively selecting from the dwarf material those plants that had longer stems and were more suitable to local conditions. Recently a group of CIMMYT scientists noted that there were farmers with some plants of the improved variety of wheat, Sonalika, which were far less susceptible to leaf rust than others. To capitalise on these developments, the CIMMYT group suggested that researchers collect rust-free heads of wheat and multiply the lines that are resistant.

The fact that farmers conduct agronomic research is borne out by the rapid spread of improved varieties of wheat in India and Bangladesh in the 1960s and 1970s. After official demonstrations were made in farmers' fields to show the potential of the new seeds, often under optimal or high input conditions, it was frequently the farmers themselves who adapted those packages to their own conditions. Since the mid-1970s, about 70 per cent of the high-yielding wheat varieties in Bangladesh have been grown without irrigation. The rapid spread of wheat has surprised many scientists and extension agents who did not appreciate its potential on the residual moisture conditions in Bengal. Unfortunately scientists have frequently overlooked the importance of farmer R & D on agronomy practices. Only too often they have seen the non-adoption of the full package as a sign of backwardness on the part of farmers or as a result of inadequacies on pricing policy, the supply of inputs, etc, instead of monitoring the creative way in which farmers have modified and adapted inappropriate packages of practices and then capitalising on such new developments by passing the information on to extension agents.

Other inventions from Bangladeshi farmers include the testing of wheat seed germination on bamboo leaves before planting to determine what quantity of seed to use; the application of fertiliser after occasional rains in winter; the use of tins, polythene bags and local drying practices to keep seeds in good condition during the monsoon; the adaptation of the rice paddle threshers for threshing wheat; the sowing of wheat on ridges (like potatoes) to facilitate irrigation on impermeable soils; the division of plots with ridges and channels to optimise the use of scarce irrigation water.[37]

These examples all serve to show that farmers create and adopt technologies. The views of many researchers in the past are demonstrably wrong on two counts: (i) that farmers *will* develop an 'optimal' agricultural system in a given environment. This environment, though, is rarely an optimal biological environment and comprises a whole set of socio-economic scenarios: and (ii) the recipient of a technology is most likely to adopt a 'correct balance' of the component parts of a technological package from a given research situation, and will also create and make changes to the technology.

Brammer (1979) makes the following observations and recommendations for researchers and extension workers:

> In order to advise farmers on improved methods of cultivation, extension staff need not depend only on the recommendations coming from Agricultural Research organisations. Research results do not apply to all kinds of land and soil, nor to all crops, and not all farmers can afford to use recommended practices. Also, in some cases, good farmers are obtaining higher yields than those produced by the 'official' recommendations.
>
> Most agricultural research in Bangladesh is being done by the farmers themselves: trying new crops or varieties; experimenting with different crop rotations or cultivation practices; testing officially recommended varieties or practices. In this way, progressive farmers find out the crops, varieties and practices which are most suitable for their land and for their economic conditions.
>
> This is not to say that 'official' recommendations for the cultivation of HYVs or the use of fertilisers should be ignored. But these recommendations should be used only where they have been demonstrated to be better than the farmers' own crops or practices. Also, many farmers do not have land suitable for HYVs, or they cannot afford to buy recommended

doses of fertilisers. In these cases, extension staff can use the best farmers' practices as the basis for making recommendations to other farmers in the area.

It is being recognised slowly that local and farm-level 'informal' R and D programmes are very important factors in developing technology which is appropriate to identified client groups. The successful dissemination of the innovations from agronomists, engineers, etc have often depended on the inputs of 'informal' R and D.

2. Grameen Bank project, Bangladesh

This project involved a credit programme aimed at assisting the very poor. The beneficiary group was identified as those with less than 0.4 acres of cultivable land and family assets not exceeding the market value of one acre of medium-quality land. Some of the most interesting work under this scheme was that carried out in relation to women belonging to rural families involved in non-agricultural activities, e.g. paddy husking, coloi crushing, grocery shop running, sweet making, etc.[38] The Grameen Bank endeavoured to change the traditional position of women by enabling them to become effective wage earners.[39]

The objectives of the policy were to (i) establish an organisational structure to provide a reasonably dependable forum through which the banking system could extend credit to those landless lacking collateral; (ii) to find a framework through which those economically and politically powerless could, by collective action and self-realisation, improve their conditions and participate in village government effectively and meaningfully; and (iii) to find a new and functional version of the cooperative.[40]

Policy implementation involved branch level staff living in the villages. The approach was informal and accessible to all. Groups were formed in villages and regular meetings held.

The results of the programme have been that new sources of income have been generated amongst poor men and women.[38, 40]

Although the new forms of employment have only, in some

cases, assisted in the wider adoption of known technologies, the benefits to the rural poor – the identified client group – have been large. It is lessons in relation to beneficiary group identification and policy effectiveness which are important here. The Grameen Bank was most specific about those whom it was attempting to assist – and it maintained a close control on credit extension and use. The scheme provides a full range of banking services and repayment rates on loans – totalling over US$1 million – have been over 99 per cent. The village level approach by bank officials enabled the project to do this. It also meant that advice and information, on what was available and how to benefit from it, were immediately accessible to local people. The results are observably successful.

III. IRRIGATION

A. GOVERNMENT AGRICULTURAL DEPARTMENTS

Irrigation in Bangladesh

Due to land constraints one way of increasing the agricultural output in Bangladesh was to expand irrigation systems and facilities. Analysis of irrigation in Bangladesh, even by hand lift pumps, highlights the very important institutional issue of structural reorganisation, i.e. it raises the wider political question of whether or not it is possible to benefit the rural poor at all under certain conditions.

In 1978 only 12 per cent of cultivated land was irrigated.[41] It was estimated that it was feasible to irrigate 20 per cent of all agricultural areas from surface water, and that most of the remainder could be irrigated by shallow tube-wells (25 per cent), and by deep turbine wells, from ground water sources.

The higher level of food output that would be the result of these irrigation proposals was important for several reasons:

(1) More food would be available for (a) the people of Bangladesh, and (b) export.

(2) Increasing levels of nutrition appear to have a negative correlation with population growth.

(3) Agricultural intensification would use some of the growing labour force.

This last point highlights the question of mechanised pumping versus hand pumping. A capital-intensive mechanised irrigation scheme requires far less labour than a man-powered scheme. In a situation where factor costs are so disparate, i.e. capital high, labour low, it seems inappropriate to recommend the former technology. Also in Bangladesh landholdings are small and often fragmented, therefore a mechanised system with a large service area would require high levels of cooperation between farmers, and is open to control by larger farmers who have (a) greater social powers and (b) easier access to credit. There is also the problem of competition for ground water during the dry season – in a match between turbine well and a shallow well, it is the deeper that will win.

However, the government of Bangladesh and aid agencies have both invested in and subsidised large-scale mechanised irrigation schemes. Edwards et al. (1978) draw out the following reasons for this:

(a) these methods were used first in West Pakistan and were transferred in the 1960s to East Pakistan with little or no analysis as to the suitability of these technical-cum-institutional packages to conditions in East Pakistan;

(b) the bias of western-trained officials and engineers towards western-style technology;

(c) the 'show' effect – i.e. the visibility of these methods to outsiders;

(d) their suitability, in terms of high foreign-exchange inputs, for foreign aid agreements;

(e) the economic bias imparted in their favour by an over-valued exchange rate;

(f) ignorance about the actual performance characteristics of these schemes.[41]

In spite of this emphasis on mechanised irrigation, manually operated networks have increased[42] as farmers found that, due to the relative costs of mechanisation and labour – the rising price of capital and the falling real agricultural wage – linked with the increasing value of rice, it was quite profitable to use manual pumps. Even so:

> It is, however, clear that the benefits of manual methods are accruing to the larger farmers because of their ownership of land and capital, and their preferential access to finance capital and agricultural inputs.
>
> If the national objectives of a fairer distribution of income and more self-reliance are to be achieved, changes in government policies are required. Institutional credit and subsidies will need to be switched from the middle-income and rich farmers to the poor farmers and the landless, if the proportion of landless is to be prevented from rising still further, and the

real agricultural wage to be prevented from falling even lower.[41]

From these comments it appears that several points should be noted in the planning of further irrigation schemes. It is important that the present social and political structure is considered in any future recommendations concerning irrigation technology or economic policy. If this approach were to be taken, it seems likely that in order to improve the situation for those rural poor who, because of past policies, are experiencing increasing landlessness and poverty, programmes will have to be aimed at structural reorganisation and job creation rather than at simply increasing agricultural output. Thomas et al. (1976) emphasise this in their analysis of public works programmes, in fifteen countries, stating:

> One must be rather sceptical about the employment creation effects of programmes where less than 35% of funds are directed to labour. Such a situation suggests that employment creation is not taking place as it should and the programme is relatively ineffective in terms of this objective.[43]

With regard to irrigation and drainage projects they noted that those immediately benefiting were those holding the land. Although significant in benefit-cost terms most cases reviewed had undesirable distributive effects.

The negative results on ownership of productive assets in relation to smaller farmers must be considered in any planning of irrigation schemes, and steps taken to avoid a situation of differentiation occurring. For the landless the only likely benefit of a scheme is in terms of wages and/or employment, thus this is an important issue when weighting the relative factor prices of a programme.

B. LOCAL NON-GOVERNMENT ORGANISATIONS

The following case study is an example of a project aimed at benefiting the growing landless sector of society in Bangladesh. It is also interesting to note the reasons why the Ministry of Agriculture supported the programme.

1. Proshika – Irrigation assets for the rural landless, Bangladesh

This study looks at an attempt by groups of landless and near-landless (with support of Proshika, a local NGO) to create and sell irrigation water to cultivators engaged in the production of HYV boro rice. The programme objectives were to relieve the problems of the landless poor, by giving them some control over the means of production, and to improve the efficiency of water use by challenging the monopoly of the large farmers.[44]

By the end of 1979 the second Five Year, and the Medium Term Production, Plans had been formulated in principle by the government. The Ministry of Agriculture and LGRD (Local Government and Rural Development) were both expressing concern over the institutional implications of the strategies and the position of the increasing number of landless. It was realised that growth and equity were interrelated and that, therefore, much of the constraint on production was social. From this recognition came the plan, whose key element was the expansion of small-scale irrigation technology. A World Bank Report in 1981 estimated that by 1985 there would be an extra 2.3 million available for rural jobs, and produced depressing figures of those already under-employed. Given this and the financial backing of the Bangladesh Agricultural Development Corporation and IDA, the plan was formulated in the hope that those landless groups who would thus participate in the production process would in turn constitute an effective demand for the expanded production.

Implementation of the programme involved the Integrated Rural Development Programme (IRDP), with the support of LGRD, preparing an experimental project through the existing Irrigation Management Cell who were already running a limited programme with landless groups. Proshika, interested in the underlying principles, presented the ideas to groups of landless labourers and received a positive response. The approach was one of placing the groups themselves in positions of researcher and beneficiary. This meant that the present variations in practice, due to variants of the social, economic and ecological systems, could be locally evaluated. This included (a) identifying relationships between the sets of variables to determine where improvements could be made to strengthen the project's financial viability, and (b) assessing the wider issues of employment, migration patterns, etc.

The success of the project was measured in financial terms of net returns, and figures of 75 per cent and 78 per cent success rates recorded for shallow tube well (STW) and low lift pump (LLP) groups respectively.[45] But, as Wood states:

> The analysis has concentrated upon the rather narrow but crucial criteria of financial viability despite the broader structural objectives of the programme. This emphasis has been intentional since financial viability is a necessary, though far from sufficient, condition for continuing and expanding this approach to agrarian reform. Both from within Proshika and by others, we have been criticised for devoting too much research, investigative and monitoring attention to the 'conventional' concern with return on investment. However, such concern is justified as an initial consideration in a programme which seeks to generate the conditions of material independence or strength among landless classes. In this sense, the net return in unavoidably an index of this material independence in the process of breaking down erstwhile relations of dependence by the landless and near landless upon landlords, moneylenders and employers.
>
> However, financial viability is by no means a sufficient condition of such independence, and may indeed disguise the truth of such relations. The measurement of 'success' in these terms can only show that, in principle, a group can honour all the new financial obligations incurred as a result of participating in rural production in this way, so that relationships of financial dependency are not actually being increased.[45]

There are a number of structural implications of the rural landless owning and deploying minor irrigation assets. Firstly, there is a noticeable process of educational and consciousness raising occurring. Secondly, conflicts of loyalty between kin and irrigation groups occur – therefore an important criterion for selection of scheme participants must be a high level of common interest. Thirdly, it was observed that where trading members dominated a group, the impact was negative. Finally, several different organisational arrangements have occurred in (i) the collecting of water charges, and (ii) those large command areas where possession by a single group would exclude others. In this case it is felt that, if a large number of groups are involved, the likelihood of private property and petty commodity perceptions is lower.

What have the benefits of this project been – and who have they been beneficial to? The landless – the client group of the project – through holding their own, have been involved in gainful employment. It is in this way that the short-term financial success or impact of the programme can be assessed. With the expansion of such schemes and increased agricultural production the secondary and tertiary benefits generated (e.g. milling for women; the manufacture, procurement and distribution of fertilisers and pesticides; mechanical services; water credit and marketing arrangements; accounting procedures; transportation, etc) in the long term could have widespread positive effects for the rural poor. The small owner-cultivator has benefited in two ways. He is now in a position of greater security through having access to water which increased the viability of his small plot; in some areas the price of irrigation water has dropped. The effects on tenants and sharecroppers have not been adequately assessed by the project to pass judgement on.

In summary then, if one gauges the success of the project in terms other than financial it would appear to have been progressive in that landless and small farmer groups have taken over the control of water services. That this has been so is mainly due to the fact that consideration was given to the social setting and the local variables involved in the issue. In relation to the former it should be noted that it was financial conditions, and the political climate (that persuaded the government to give their backing to this plan) that has given rise to structural changes.

2. UNICEF/NGO Water Project Steering Committee, Kenya

The following study of the Olosho-Oibar district of Kenya has been included because it illustrates several points. Firstly, the beneficiary group and the programme objectives regarding this group were not only clarified at the beginning, but also all the way through the programme. The technological answers were a result of an analytical procedure that looked carefully at all the variables affecting the identified group.

Although loosely attributed to this particular category the projects described here are of a mixed nature. The linking factor has been the UNICEF/NGO Water Project Steering Committee's coverage and assistance.

In Kenya women's cooperatives or Harambee groups are traditional. The UNICEF/NGO Water for Health Programme identified 15 primary projects to support and to work with as a cohesive and strong basis. The only selection criterion was that the programmes had to be initiated by women and should have the objectives of (a) improving women's participation in rural development, and (b) alleviating some of the problems of daily life. Thus a whole spectrum of activities and interests were covered from basic irrigation services to those of health, hygiene and education.

The Olosho-Oibar district had suffered from a three-year drought. Many animals had died, thus causing a shortage of both meat and milk – both previously important parts of the Masai diet. Women were travelling at least five miles a day in search of clean water, using time that otherwise would be spent in income generating activities – the revenue from which could be used to procure nutritional supplements (e.g. beans, maize) for the family:

> With the help of government experts, a water source was tapped in the nearby hills, which enabled the Committee to supply 13 Masai homesteads along the valley with clean water, collected in conservation tanks donated by the Zonta Club and the Kenya Association of University Women. Prior to involvement in the area, the women had managed to save K.shs 600 for this project from the sale of handicrafts. The Committee's interest in their work attracted the attention of the men in the area, who donated K.shs 5,000, which was used for the purchase of pipes.[46]

While dealing with the subject of water pipes an interesting example should be noted:

At a recent conference in London a speaker told the story of participants in a mission to a developing country who were so disturbed at the conditions in a village they visited that they decided, on their return home, to help the people there by providing them with a piped water supply. In due course the necessary piping was sent out but, when the technicians followed to install it, they discovered that it had been used to make benches for the men's meeting house. There was no need, the men said, for a piped water supply; what would the women do all day if they did not have to fetch and carry water? It is obvious that the women's views would have been very different from their husbands', but at no time do they appear to have been consulted, either by the men or by their would-be helpers.[47]

Again this highlights the institutional responsibility to properly define the beneficiary group, and to work specifically with them. Following the installation of a clean water source to the Masai homesteads of Olosho-Oibar the UNICEF/NGO Committee requested that the Family Planning Association of Kenya, the Co-operative Education office and the Ministry of Health take an active interest in the area. They were specifically asked to help in the dissemination of information regarding cleanliness and hygiene. The groups agreed and a full-time nurse was posted to the region. Regular educational sessions were also established, and under the sponsorship of the Associated Country Women of the World two local women were trained to work with the people in the area.

The results of this particular project have so far become manifest in that scabies, previously common among the children, has almost disappeared; latrines have been built by the community – as has a clinic and house for the nurses; a communal vegetable garden, growing cash crops, has been established; and the marketing network has been improved for the sale of the women's beadwork.

The apparent success of the programme can be attributed to several factors. As stated earlier, the beneficiary group were correctly defined, and used as a source of both information and labour, i.e. the importance of local rather than outside involvement was recognised. The existing cooperative system was used as the basis for development – blending the traditional and the new inputs. The committee provided the initial thrust for change in identifying the need for and installing a water supply – but from that point their role became that of observer and information bank. The onus was on the local people who were able to turn to the committee if a need was felt. The committee then provided links with other institutions and agencies who were able to carry out the functions to which they were suited. It is an interesting example of how inter-agency links can be utilised to form a cohesive and progressive programme.

3. Gram Gourav Pratisthan (GGP), India

The following is a case study of the growth and success of a local NGO. Its effectiveness, as is shown, can be attributed to the sense of scale the coordinators maintained, especially in relation to the importance attached to knowledge of a specific locality.

It was the 1971-1972 drought and consequent famine that initiated the funding of the Gram Gourav Pratisthan (Village Pride Trust or GGP) in the Pune district by a wealthy local man and his social worker wife. The need for such an organisation was discovered when the couple, the Sahlunkes, visiting drought relief projects in the area, found villagers involved in temporary and often useless work. The people themselves knew this, and expressed their frustration to the Sahlunkes. They felt that work involving the creation of percolation tanks and check dams would be of far more use.

By 1974 the charitable trust GGP had been established to raise funds from governments and other contributors. Their objectives were to build small-scale irrigation works on sites that would benefit small-scale landholders, and would utilise local labour. 'Water Committees' were organised at village level to raise money for the schemes. Generally the villagers themselves chose the sites for the irrigation works, and approached the Sahlunkes for advice on funding and technology. In the beginning GGP supplied 40 per cent of the money, the government supplied 40 per cent and the target group of disadvantaged farmers raised the remaining 20 per cent. In 1981 the government withdrew their support; money is at present being donated by the Dutch and Swiss Governments, a church group and some Indian industrialists.

Many of the reasons for the government's backing out of the scheme are political. GGP farmers have tended to locate wells close to irrigation channels and have been involved in several court cases on charges of illegal appropriation of groundwater. Water ownership and management has been collective, and as in the Proshika case, a sense of community and group consciousness has begun to emerge amongst these, usually lower caste, farmers. Many problems have been encountered as a result of these unpopular social developments. Electrically driven pumps are used in some of the more ambitious projects[48], and connection to supplies has been unnecessarily delayed.

Despite these setbacks GGP projects are having a noticeably positive effect on the poorer groups. A 300 per cent increase in agricultural production has been recorded on lands irrigated under these schemes.[49] Vegetables are double cropped and there has been an increase in the amount of wheat grown.

Because of this success, pressure has been exerted upon GGP to extend their interests beyond the Purandhar Block, but as Sahlunke states:

> Our goal is not to make this a big organisation but to give aid and assistance to those who want to replicate our efforts. My whole approach is to make the block the operational unit. Every institution you want to build in India should be at the block level.

To assist in this the Ford Foundation have donated $145,000 which will be used to build a development centre. The role of this centre will be to educate rural school-leavers in water management, crop planning, accountancy and marketing. The trainees will then assume the position of field workers, and it is projected that they will assist in the execution of 500 further projects, irrigating 25,000 acres of land and hopefully providing full-time employment for 25,000 families.

The impact of GGP projects is self-evident. Their success is partly to do with the sense of scale that has been maintained and partly due to the emphasis placed on local involvement. Regarding scale, as Sahlunke stated, there is irrefutable evidence to support arguments as to the efficiency of small units such as the programmes with which we are concerned. Management and technological problems are less to begin with, especially where water is dealt with on a cooperative basis. This cooperative organisation seems to have worked in two ways. Firstly people have felt that the situation is one of consent rather than coercion, i.e. the farmers feel they have control over this particular factor of the production process. Secondly the rural poor have begun to identify themselves as a group with common interests – interests that are best served by working together.

Institutionally, agencies can benefit from this local NGO experience by noting (a) the need was specified by local people who were then involved in programme formulation and practice, and (b) the parent group remained small and flexible, providing information and expertise when required, acting as a link with other funding bodies and agencies, and providing the impetus for the development of similar groups elsewhere.

4. Utooni Development Project (UDP), Kenya

The Kenyan tradition of cooperative work has assisted greatly in the success of this project, planned and organised by elected committees of the Kamba self-help group, and it is useful to note the way in which the UDP planners utilised this system.

A number of problems have been tackled, the most important being initially identified as that of the provision of clean water. Having only 70mm of rainfall per year, and with the bulk of this falling in a short space of time, the UDP had to solve the problems of both water collection and storage. The village had, in the past, relied on an unprotected spring which was open to pollution from both the animals and humans that used it. This became the primary target in the quest for clean water. The spring has been protected and its water piped to a storage well, from which it is withdrawn by a handpump situated on a concrete platform. Construction of a small animal-proofed earthdam then took place below the spring source, creating a reservoir. Water is conveyed through the dam via a small pipe, then through a filtration trench and thus to the storage well. A concrete arch dam has been constructed across a local catchment area to store rainwater that is normally lost as run-off. The UDP committee also plan to build three sand dams in selected sites around the village. These dams store water in the space between particles of coarse sand. Other developments here included rainwater harvesting and storage of the water in locally constructed ferro-cement jars, digging of a shallow well for irrigation of a tree nursery, and the construction of pit latrines. The 'Ghala' tank idea originated in Thailand when UNICEF observed rural people carrying water in baskets covered with an impermeable layer of tree resin. The idea is interesting in that it demonstrates the adaptation of an old technique to a new technology – the result of which is the ferro-cement storage jar.

The success of the project can be measured in that clean water is now provided for most people in Kola. The pump is located fairly centrally so the benefits to women, previously travelling often quite long distances to collect the muddy polluted water from the spring, are particularly noticeable. The project has not been without its problems – time and finance are both scarce factors in Kola, but the increase in the standard of living – which included a decrease in the incidence of disease – has been noticeable and has engendered further enthusiasm from the participants.

An evaluation of the project isolates several variables that have enabled its progression and continuing achievements; (i) the village has a strong traditional collective ability, every household contributes both labour and money; (ii) because of the local components and influences in the committee there is an instinctive understanding of need, problems and possible solutions; (iii) the organisational structure, i.e. village self-help group steering committee, ensures that every participant has access to the decision-making process; (iv) there is a continuing awareness by the group of the strength of its collective nature and the effectiveness of its actions. In terms of the technology used, especially in relation to the construction of the ferro-cement storage jars, care has been taken to use:

. . . easily-available materials, needing simple skills and simple equipment, and requiring simple organisation of labour. Sand, cement and wire are available in most areas and are familiar materials to most people. The raw materials are easily transportable, only cement requiring protection fom the weather. The construction techniques are simple and generally require little trained supervision. No sophisticated tools or machinery are needed and maintenance or repairs are not complex.[50]

The breadth of the Utooni Development Project is large. It has analysed, and solved technically, one problem. However, this has not been done in isolation: the repercussions and possible spread effects of action have been carefully monitored, evaluated and capitalised upon (e.g. seepage from the filtration trench has been used to irrigate banana trees planted along its edge). Educationally, health programmes have been set up and villagers have been sent to Nairobi to take part in the UNICEF training courses set up by their Appropriate Technology Projects Office. These people then return to the local area and are able to disseminate the information. Thus the committee has begun to establish links with other institutions and agencies that could be of assistance to the project.

In summary then, the important issues are once again proved to be those of locally specific analysis of problems; an awareness of the cause and effect relationships between all the variables of that given situation; an inclusion of all participants in decision making, information gathering and programme implementation; and finally, of great importance, taking the existing social, economic and technological environment, extrapolating those factors considered of value and utilising them to form a dynamic synthesis of old and new.

C. INTERNATIONAL AGENCIES

Oxfam, Somalia

Continuing hostilities in the western area of Somalia have led to a severe refugee problem. The problem is exacerbated by the three year drought that the region has been subject to, and the situation has become such that immediate relief in the form of water supplies is necessary. Case studies of disaster relief are somewhat different from those generally used in this paper – but this example does illustrate the point that engineering hardware can be effective and appropriate, especially to an emergency situation, and that agencies have a responsibility in such situations that must be recognised.

The technology introduced took the form of permanent wells 4-6m deep – hand dug by the refugees. A petrol-driven rock-breaker was used in hard rock areas. The siting of the wells, under an English engineer's supervision, was on the banks of the dry sand-filled river beds. This was so they would be safe during the flood season. A number of small temporary wells were dug in the actual river bed for supplementary use during the dry season. The main wells were lined with corrugated galvanised steel, perforated by hammer, chisel or welding rod burns, below the water line. These were then backfilled with roughly graded rocks, stones and sand. So far, the hardware inputs were relatively easily obtainable and maintainable. The decision was made to utilise solar pumps, with a small number of diesel and petrol engines for back up during times of peak demand. The latter, as was expected, have problems regarding maintenance, spare parts and clean fuel availability. The solar pumps, donated by the German 'Freedom from Hunger' organisation, convert solar radiation directly into electrical power by means of a photovoltaic (solar cell) array. They suffer from none of the above problems of a diesel or petrol power source, and their working life is long. However, if the pumps or cells went wrong they had to be sent back to the manufacturer. Generally, though, these pumps have proved efficient and relatively reliable.

What has been the effect of this emergency technological solution? In terms of quantity, water is now more readily and widely available. Regarding quality of supply the difference is particularly noticeable in the decline of typhoid and dysentery during the wet season. Before the advent of these wells the refugee population obtained their water by scooping it from the saturated sand of the river bed. During the dry season, due to the filtration effect of the sand, water gathered in this way was fairly clean. However, during the rains when the water table was nearer the surface supplies became polluted from runoff from the camp areas, giving rise to a high incidence of disease.

Initial indications are that effects of this emergency programme have been positive although not enough time has elapsed for a realistic evaluation of the situation. The next phase is that of 'consolidation' – i.e., one where 'emergency' supplies will become 'permanent' supplies. How this is to be done is at present under consideration – and proving to pose problems in the stress it would place on groundwater, human and financial resources.

Institutionally what can be learnt from this case is that agencies play an important role in disaster relief, and the effectiveness of this relief relates directly to the appropriateness of its nature to a given situation. In the next stage, that referred to here as the 'consolidation' stage, it is important that the agency does not become 'locked into' the technology type solution that was used as an emergency measure. It is then that the agency role, policies and responsibilities must be most sharply evaluated and defined.

D. INFORMAL RESEARCH AND DEVELOPMENT

The bamboo tubewell, Bihar, India

This particular example is of interest, as it provides a case where indigenous developments have come in response to and out of an externally introduced technology.

The late 1950s saw an attempt by the Indian Government to increase agricultural production in the Kosi area by the introduction of tubewell irrigation. The wells were constructed from 10 to 15cm iron casings with brass screens and sunk by peripatetic rigs to a depth of 45m or more. Transport costs of these rigs were high, pumps were powered by electricity (posing problems to the Bihar State Electricity Board of linking the scattered pumpsets) and water was delivered through a system of brick or concrete channels. Initial response to the technology was limited mainly to a few large landowners. The 1960s 'green revolution' package, coinciding with a period of drought, included a campaign involving direct government credit for the sinking of tubewells.

The results for some farmers were most profitable – but the wider implication of this policy was one of increasing rural differentiation.

However, the benefits of irrigation were recognised and farmers embarked on a series of experiments aimed at reducing the costs of tubewells:

> A few contractors and farmers with business contacts outside the region found that they could substantially reduce investment costs by importing their own materials, sinking shallower wells and using local contractors to install the wells. By privately installing wells, they also cut out the lengthy delays and other hidden costs of credit-financed investments.[51]

It was also discovered that iron screens could be used to replace the expensive brass screens, and that drilling could be carried out by using a simpler and cheaper 'sludger' method. A mobile single diesel engine mounted on a bullock cart powering several local wells was found to be more economical than the government-recommended single site electrical pump. Earth channels replaced the planned brick and concrete networks.

The bamboo tubewell was another direct result of this type of informal research. Its development had taken the form of coconut coir wrapped around steel tubes and, later, bamboo frames. Finally a well was successfully sunk using a bamboo casing and a coir and bamboo screen.

The impact of this indigenously developed technology was high, not only to large and small farmers but also for landless labourers:

> The techniques of assembly and sinking of bamboo tubewells as well as the complementary work on land levelling and channel construction largely involved unskilled labour and a minimum of capital equipment. My estimate is that in 1972-73, 300,000 man-days of additional employment were created by the fabrication and sinking of at least 14,000 bamboo wells, and 100,000 man-days through subsequent earthworks. In addition, the maintenance of a stock of 150,000 to 200,000 man-days of employment annually, according to whether one assumes an average life expectancy for wells of 4 or 3 years. However, this is considerably less than the additional employment generated by tubewell-irrigated

farming when even in 1971-72 it was estimated as at least 1.7 million man-days when there were less than 5,000 operational wells in the region. Approximately 30 per cent of incremental net product from more intensive cultivation went to agricultural labour. The low cost of bamboo tubewells and the development of a market in pumpset services enabled many more farmers to introduce tubewell-irrigated farming profitably. It was the overall labour-using character of the package of innovations associated with tubewell-irrigated farming more than the labour-intensive nature of bamboo tubewell fabrication and sinking techniques that had the greatest impact on employment.[51]

Large farmers were able to sink a number of wells that utilised a single pumpset. In 1972 subsidised credit and the development of a pumpset service market improved small farmers' access to the technology. However, it must be noted that the sharing or marketing of pumpsets introduces problems of organisation and timing. Credit costs also remain high for the poor farmers, a factor reflected in their access to other inputs, e.g. fertiliser, seed, pesticides, herbicides and mechanical power during periods of peak labour requirements.

However, despite these remaining issues, the diffusion of bamboo tubewells is interesting for several reasons. It has assisted in the spread of irrigation, and thus to some extent has provided employment and higher incomes for labourers in the agricultural sector. The institutional lessons are those to be drawn from looking at how and why this technology developed from the introduction of another. It was developed locally by those who needed it (whether the well would be viable in areas other than Kosi, an area of coarse, sandy alluvia with a high water table, is another question). Because of this, technology produced was appropriate to the physical, social and economic environments, and it has thus served to illustrate the potential that exists for the adaptation of externally devised hard or software. It is significant to note that the local engineers and the minor irrigation department were originally sceptical and hostile to this technology when it first started to spread. It was interest by the Kosi Area Development Commissioner which helped increase the rate of diffusion of this technology.[52]

IV. POST-HARVEST TECHNOLOGIES

A. UNIVERSITIES AND RESEARCH ORGANISATIONS

The need for investigation into Harvest and Post-Harvest Storage

In spite of Farming Systems Research little attention has been paid to Harvest and Post Harvest Storage (HPHS) of crops. Where work has been carried out consideration has been given mainly to the crop specialisations of the larger research institutions, e.g. rice, maize or wheat.

The following discussion attempts to show:

(a) the reasons why Harvest and Post Harvest Storage research is considered necessary;

(b) what particular crops should be concentrated upon given the interest here in small farmers;

(c) the value in the 'bringing together' of modern and traditional technologies; and

(d) the problems which exist in this area of technology when the development and promotion of certain techniques can lead to the greater impoverishment of very poor groups (especially women) in rural areas.[53, 54, 55]

With increasing emphasis placed on the family farm as the centre of research and development activities there is a spreading recognition of the need for HPHS research. Maxwell (1982) argues for the importance of this issue in labour-scarce situations on the basis of five propositions:

1. The HPHS accounts for a major share of the labour required for crop production; this labour is needed at a busy time of the year and thus has a high opportunity cost.
2. The requirement for cash is often too high to hire labour, contract threshing machines or transport produce. Cash is also particularly scarce during harvest, with a high opportunity cost.

3. The cumulative effect of losses at different stages of the HPHS can be an important hidden 'cost', as also can quality deterioration, especially during storage.

4. Harvest and post-harvest technologies affect marketing strategies, which in turn affect the price obtained and thus profitability.

5. Appropriate adjustments to the HPHS can be devised which save on scarce resources or otherwise increase income.

These important points are illustrated in this case study of Colonist Farmers in Santa Cruz, Bolivia:

> The study area is comprised of one segment of the ring of colonisation that stretches around the Amazon basin from Brazil to Venezuela and that contains around 15 million people (Barbira-Scazzocchio 1980). It is a subtropical area some 160km square that has been settled from the original high forest over the past 20 years; it contains some 15,000 families or about 22 per cent of the rural population of Santa Cruz Department living in nine main colonisation areas that have received varying degrees of government support. Typical farm size ranges from 20-52ha and agriculture is based on slash and burn techniques with upland (unirrigated) rice as the main cash and subsistence crop.[56]

The current agricultural situation of the area is one referred to as the '*barbecho* crisis'. *Barbecho* is the forest re-growth occurring when a slash burn cultivator moves from his initial site. However, due to land constraints the farmer will return to the same spot before the forest has time to regenerate. Yields are consequently lower and weeding takes longer. This downward spiral can be halted by de-stumping and mechanisation, the development of livestock enterprises, crop diversification and new techniques. However, it has been noted[57] that this tends to lead to increasing social differentiation, labour bottlenecks (especially at harvest time), marketing problems and a fall in real farm profits.[58] One issue to be considered is that improvements to HPHS may be necessary to facilitate adaptation of technological innovation in other areas of the production process.

The value of HPHS research is demonstrated in this case as it shows that HPHS accounts for 50-90 per cent of total expenditure before labour is hired.[59] No land figures are available for the economic losses of present HPHS techniques but they are estimated to be high:

> Preliminary research shows that the percentage of grains damaged after twelve weeks in traditional storage reaches 6% for maize and 9% for rice.[58]

Thus marketing is highlighted as an area of importance, as the farmers of Santa Cruz tend to involve themselves in this as soon as possible after harvest, (a) to minimise storage losses, and (b) to meet the cash cost of hiring harvest labour.

This all serves to emphasise that there is both room, and a need, for HPHS research in the FSR framework. This area is currently one whose true worth is unrecognised and consequently has little input in terms of manpower or finance. Agencies therefore have a role to play in the investigation of this area especially concerning those crops grown for risk avoidance and subsistence such as cassava, yam, taro, potato and sweet potato. These farmers are those most at risk and HPHS losses are more significant to them than to the richer farmers; it is also suggested that due to the larger research institutions' existing crop interests, these crops are unlikely to be given much attention.

Coursey (1982) discusses this issue, emphasising the importance of both these particular types of crop and the local technologies used for storing them:

> Studies in the area of post-harvest technology have so far mainly concentrated on grains and other durable products which are stored dry, usually at moisture contents below about 12 per cent. In these products, post-harvest deterioration is largely caused by attack of external agents such as insects, moulds or rodents and not from endogenous factors. Most of the lesser amount of work undertaken on perishable crops has concentrated on high-unit-cost horticultural products such as fruits and vegetables rather than on the low-unit-cost staple foods such as root crops. Different approaches are therefore necessary when dealing with tropical root crop products and in many cases traditional technologies, developed in the distant past within subsistence agricultural societies, may be especially appropriate. During the last decade or so, much effort has been devoted by bilateral and multilateral aid agencies to research and development under designations such as 'rural tech-

nology', 'grass-roots technology', 'small-scale technology', 'intermediate technology' or 'appropriate technology'. However, most of the conceptual philosophy of such work has been derived primarily from the conventional scientific approaches to the developed world, and has neglected the very considerable corpus of traditional knowledge relating to tropical crops that has been accumulated over centuries or millenia within the societies that grow and use them. Indeed, there has often been a tendency among those who have received a modern scientific education to reject traditional technologies as 'primitive' and fit only to be displaced by sophisticated modern systems, even though the latter may sometimes represent sub-optimal technologies for the situation. In the case of the tropical root crops, the neglect of the traditional wisdom is especially unfortunate, as the underlying philosophies of the cultures in which they are grown are extremely alien to those of Europe, within which scientific thinking developed.[60]

This store of traditional knowledge, especially of the post-harvest technology of tropical root crops, has remained largely untapped but possibilities nevertheless exist for the interaction of modern scientific concepts with these traditional systems.[61]

A primary crop, in terms of amount produced, in the developing world is cassava. It is interesting in that the swollen roots for which it is grown serve no biological function, i.e. they are not organs to support the plant through a period of dormancy. Once detached from the plant they do not store well – a problem overcome traditionally by either drying the roots or leaving them in the ground until needed. However, with increasing pressure on land this method is becoming non-viable. The Amerindians of Amazonia have developed storage pits. Observing this success in prolonging the out-of-ground storage life of the root for several months, investigations were carried out by natural scientists (Booth, 1974; Booth and Coursey, 1974). They discovered that the environmental conditions established in the pits favoured wound healing – the forming of a wound periderm which inhibits endogenous physiological breakdown and secondary pathogenic invasion. As Coursey states:

This interaction of modern scientific/technological investigations with studies of ancient traditional storage practices is now leading to the development of improved storage and transportation systems, as it is now understood that the avoidance of water stress is essential to the long-term preservation of cassava roots.[62]

The value of traditional developments is being increasingly recognised by research workers. Cassava as a toxic root must be processed before ingestion – and many of the traditional practices are now being looked to for the best basis of modern technological processing. For example in Brazil *farinha de mandioca* is produced at an industrial level, but the process involved is no more than a scaling up or rationalisation of the traditional process, i.e. using hydraulic presses instead of the traditional *tipiti* (woven wicker press). An interesting comparison is that of the transfer of this technology to West Africa, where adoption rates have been low. Analysis shows that economic reasons lie behind the apparent failure of the new technology, i.e. the cost of labour used in traditional methods is far less than the cost of the technology.

Two lessons then from this example, given that the importance of HPHS is accepted. (1) There is a valuable role to be played by research and development agencies in the synthesis of traditional and modern technologies. Much of the work can only, therefore, be carried out in the farmers' fields where information exchange between researcher and producer has to be good. (2) The second point relates to this emphasis on on-farm research. The researcher must understand the complete decision-making environment of the producer. Not only will he then be able to assess traditional methods in physiological terms, but hopefully he will also be able to ascertain the value in changing or utilising the technology observed in that particular situation.

The International Potato Centre (below) has been conducting research very much along these lines and with apparently successful results.

CIP (International Potato Centre), Peru

CIP is one of the exceptional international agricultural research institutes where anthropologists, economists and natural scientists are combined together in the Centre's major programmes. It is possible that CIP, being one of the most recent of the international institutes, was able to learn from the institutional lessons of its elders, as regards the need for, and difficulties of, ensuring interdisciplinary analysis. The Centre has a strong bias towards working on the problems of resource-poor farmers[63] and their commitment to starting and ending with farmers in the research and development process is typified by their 'farmer-back-to-farmer' methodology.[64] The importance of, and respect for, poor farmers is also demonstrated by some of their publications.[65] The following discussions illustrate client-oriented research in CIP's programme.

(a) PHILIPPINES

Seed potato storage problems stem from the fact that the potato is a vegetable tuber and does not store as easily as grain: it is highly perishable. Inadequate storage – improper layering or stacking, or inappropriate light or ventilation – leads to both pathological and physiological losses. Seed potatoes stored in the dark produce long, weak sprouting shoots which tend to break easily, are generally not suitable for planting and have to be broken off by hand prior to planting. Premature sprouting weakens the seed tuber and this leads to yield reductions.

In addition, studies in Peru indicate that national seed distribution programmes for certified seed provide a very varied quality product (varied in terms of yield per tuber) at a fixed price and generally no better than good farmers' own seed. This emphasises the importance of the farmers' own seed. Seed is also a major component of production costs, and it is widely believed that the increased utilisation of healthy seed could contribute greatly to improved farm income through increased yield per tuber seed.[66]

CIP have therefore been involved with the CGIAR centres in an inter-disciplinary attempt to produce a technology suitable for the solution of this problem.

The client group for the programme was wide – highland potato farmers in general, but it was felt by the research team that benefits should accrue particularly to the small farmer.

The study revealed that:

ware (consumption) potatoes for sale were generally sold at or very soon after harvest and that it was common for producers to store seed and ware potatoes together.

Deterioration of potatoes in storage was rarely seen by the farmers as a complete loss because of subsequent utilisation of deteriorated potatoes which could be processed (naturally freeze-dried *chuno* or *papa seca*) for later consumption or fed to livestock, notably pigs. Nevertheless, limited supplies of adequate seed were a constraint to greater yields and better income.

Research trials were initiated using a well-known seed potato storage principle – that of diffuse light to control sprouting – that was common practice in developed countries prior to the spread of cool storage facilities. These trials were started on an experimental station in the Mantaro Valley.

A basic understanding of actual production processes and potato storage systems was essential to this research process. All farmers in the region were found to store part of their crop. The proportion of the harvest stored by the farmer varied inversely with the amount of land in potatoes. The most common storage systems were in the dwelling houses, either on the ground or in the rafters or attic. Outbuildings, generally multipurpose buildings, were used by some of the larger farmers. Field storage was found to be a declining option due largely to the problems of theft. It was noted that storage was principally for seed and consumption and that seasonal price changes rarely compensated sufficiently for storage costs including commodity deterioration.

From this a number of guidelines were synthesised for designing the research effort on improved storage systems; since the majority store in their homes, this would constitute the area of concentration; outbuildings for larger producers should also be considered; field storage, despite its declining popularity, should not be neglected; improved storage systems should be built from local materials, be cheap and multipurpose, and seed potato storage should be concentrated on.[66]

In the Philippines initially, after the identification of the dark-induced sprout growth problem, research station experiments were set up. Successful results led to the technology being taken out to the environment for which it was being created, i.e. the farm. The main problem which arose was that of cost and supply of seed storage trays – but due to the excellent two-way dialogue that had been established between researcher and farmer this was one promptly identified and dealt with. A second run of experiments proved that the storage units (set up under the eaves at the back of houses) and the accompanying practices could indeed solve the problem.

The extension work included farmers' meetings, courses and local demonstration facilities. It was found that certain farmers adopted the whole package while others were selective in their use of practices, incorporating them into their old storage facilities:

> It was found that early adopters were nearer to the demonstration sites, they stored potatoes longer, cultivated larger than average areas, were better educated, more likely to be members of local seed growers' associations, and more likely to be participants in the Seed Production Programme than the late or non-adopters.[67]

From these experiences the following lessons were drawn:

1. The close links with the farmer cooperators throughout the research process gave the effort the flexibility and responsiveness to enable it to adapt readily to focus on the essence of farmer problems: integrated within this was the interdisciplinary approach of the research team.

2. The development of a simple technology which could, in this case, be adopted and adapted, both in part and as a whole, to farmer conditions and requirements. By concentrating on small farm/household level technology, the technology was widely acceptable across a range of households, but especially to the small farmer.

3. The importance of field demonstrations in the diffusion process.[67]

The next case study adds strength to the argument that those issues outlined above are essential to institutions involved in this kind of research.

(b) PERU

Rhoades and Booth (1982) found similar patterns and rates of adoption in a similar experiment in the Andes. Again the technical thrust of the programme was one of improving storage, therefore decreasing losses, for the hill farmers of the Central Andes.

The mechanics of the study and the workings of the interdisciplinary team are interesting in themselves. At first the physical scientists remained on the regional research station, while the social scientists set out into the field. This led to an intra-team debate over the term 'storage losses'. The anthropologist discovered that:

> Central Andean farmers did not necessarily perceive small, shrivelled or spoiled potatoes as 'losses' or 'waste' (Werge). His evidence revealed that *all* potatoes were utilised by farm families in some form. Potatoes that could not be sold, used for 'seed' or immediate home consumption were fed to the animals, mainly pigs, or processed into dehydrated potatoes (e.g. *chuno, papa seca*) storable for as long as two to three years. In addition, some wives informed him that in culinary quality the shrivelled, partially spoiled potatoes were sometimes preferred.
>
> These observations were, as one of the biological scientists put it, 'the beginning of understanding a reality: namely that we scientists often perceive technical problems through different eyes than farmers. Losses to us were not necessarily losses to farmers.[68]

However, discussions led to the discovery that 'losses' were, in fact, incurred – not in terms of 'wasted' potatoes, but in terms of 'wasted' time and labour that was consumed in the 'de-sprouting' of the long shoots produced by seed stored in the dark.

The team were then able to narrowly identify the problems – the need to decrease sprout length and improve the quality of the tuber. Use of the diffused light principle was one solution for the improved storage of seed potatoes (greening occurs, therefore tubers are rendered inedible and unmarketable and only suitable for seed). The storage facilities designed at the research station were modified by farm trials to incorporate the already existing structure of the Andean farmhouse. Simple collapsible shelves made from local rough timber were designed to be stacked under the verandahs. The cost was low and the materials were familiar.

Patterns of adoption were very similar to those of the preceding case in the Philippines:

> Investigations of farmer response revealed that in both countries the new technology was continuously being refined and altered by farmers. In other words, adaptive research – but this time almost exclusively through the initiative of the farmer – continued well after the scientific team had finished its major adaptation and testing activities. Farmers rarely copied exactly the prototype store designs. They blended the new ideas with local architecture and, if a new store was constructed, made changes to suit their own concepts of space and design. Farmers were proud of these changes and the CIP post-harvest team is convinced farmers will be more likely to accept changes if they actively participate in this final research process.[68]

It was from these CIP experiences that the 'farmer-back-to-farmer' model was formulated. The principle of it is that for agricultural research to be successful it must *begin* and *end* with the farmer.

B. LOCAL NON-GOVERNMENT ORGANISATIONS

1. Eastern Technical Institute (ETI), Sri Lanka

Post harvest treatment of manioc

ETI have been involved in technical training since 1960. This case follows the first attempt by them to introduce a technology aimed at the direct creation of jobs as cheaply, simply and effectively as possible. Rather than the process itself, it is the effect and mechanics of implementation that are of value to this paper: as it is an attempt to develop an appropriate technology that links production of a survival crop to the cash economy.

The programme was initiated in Batticaloa which is in the dry zone of Sri Lanka. Having the attribute of being able to grow without irrigation, 2,000 acres of manioc were planted annually in this area.[69] Most is grown by slash and burn methods on the catchment area of the ancient irrigation tanks – many of which are at present being restored:

> There is a crucial social division between those who labour for them, eking out extra food and income from cultivation of the wild, dry catchments. The former group naturally have command of development resources . . .[69]

It was the latter that ETI identified as their beneficiary group. Casual labour was available to them on the maintenance of irrigation works or, seasonally, on the paddy harvest, and although little absolute unemployment was recorded, living conditions and rural incomes were low.

The process is one of extracting starch from the manioc by peeling, grating, washing, settling and solar-drying the root. This involves a rasper (made from offcuts from a teak mill and used drums from a paper factory), a washing tank (40-gallon drum – waste product of the transport system) and the provision of water to the production site (2 gallons for every pound of starch produced). All equipment is moveable by bullock cart – an important factor as the unit must be moved three or four times during a year to ensure supplies of manioc and water. The dried starch is transported once a month to the centre at Batticaloa where it is ground and sieved.

The scheme is apparently successful in that jobs for casual labourers and small cultivators have been created and there is now a reliable local market for a survival crop that gives small-scale producers access to a cash income. Production is mainly self-managed:

> Each of the . . . units is run by a *Pullami* man, each tending to compete with the other to consume available manioc supplies, sometimes arranging shiftwork around the clock. The units are paid only for starch produced at a rate per pound agreed yearly . . . payments are made only to unit managers, and checks indicate that proceeds are distributed equitably.[69]

This rural, labour-intensive industry has managed to compete with more mechanised producers in the marketing of this product, derived from a traditional root crop. Again this success story has the same underlying factors as other favourable projects. The objectives and beneficiary group were clearly defined at the outset of the programme, and adhered to at all stages of implementation. The technology was the result of innovative thinking and a response to a local environment, i.e. easily available resources were utilised from the manioc roots, through the actual technological hardware, to the labour available. Emphasis was placed on developing local capabilities and giving responsibility to the producers. Consequently these workers feel they have control over their product and are apparently willing therefore to commit themselves to the programme.

2. Comilla Cooperative, Bangladesh

Manually operated wheat threshing machine

This case study is of one project carried out under the programme of the Bangladesh Academy for Rural Development and of Voluntary Services Overseas (VSO). It has been selected for the way in which it shows how a project can move away from its proposed doctrines. That this was done intentionally is not implied, but as the case shows, due to the inability of the team to identify explicitly and maintain a constant working relationship with their intended beneficiary group, the research and development work progressed in such a way that by the end of the programme the technology produced was completely inappropriate and at worst might serve to exacerbate the existing problems of the original beneficiary group.

The Comilla Cooperative Karkhana began development work at the end of 1979 on a man-powered threshing machine for wheat. The work came about in response to what is described as 'heavy demand from Bangladesh farmers',[70] who having recently begun to grow wheat were finding threshing a problem:

The idea was to introduce the principle of a rasp bar threshing drum and concave, used on European machines, as the threshing mechanism most suitable for wheat. The basic operation of the rasp bar threshing machine is as follows: wheat is fed into the machine between the rotating drum and the concave; the rubbing and beating actions of the rasp bars attached to the rotating drum and the parallel beater bars on the stationary concave cause the separation of the grain from the straw.

The smallest thresher designed to date (that is, when the Karkhana began development work) was the U.K. National Institute of Agricultural Engineering (NIAE) model, which had a 2½-hp engine. To design a manually powered machine, several parameters had to be considered.[70]

These were (i) the power input; (ii) the drum speed; and (iii) the concave.

The first prototype, using as many parts of existing threshers as was possible, gave an output of between 6 and 10 kg/hr, depending on the variety of wheat used. It took three operators to achieve these figures, and problems were experienced of heads breaking off and passing through the machine unthreshed. Modifications were made, including some to the concave: increasing the number of beater bars, and enlarging the feed entrance. It was noted that for the optimal drum rotation pedal speed was too high for the operators.

In spite of this:

There was great pressure from local farmers and businessmen who wanted to buy the threshing machine. When initial tests looked favourable the Karkhana manufactured and sold eleven machines which definitely proved the existence of a market for a manually operated machine which was simple to understand and operate, and within the price range of a middle-class farmer owning two to three acres in Bangladesh. The sale of these machines also gave the Karkhana a good opportunity to observe the machine in use in the field and to look out for any defects.

The main objection from users was that the thresher's output was not high enough and too many operators were required: output per man-hour was not competitive with traditional threshing by beating (20 kg/man-hour). Thus, the Karkhana was greatly encouraged by the market demand but saw a need to further improve the machine's performance.[70]

It should be noted here that between 30 and 50 per cent of rural households in Bangladesh may be landless and that agricultural employment may be the only significant source of income for landless labourers and farmers with very small holdings.

The second prototype involved further changes to the drum diameter, concave size, gearbox, pedals and frame. Its performance was worse than the initial model. It was therefore concluded that human power was not sufficient to operate a rasp bar threshing mechanism for wheat.

The third prototype involved the fitting of a 1-hp electric motor to the first prototype. This meant that only those farmers who (a) had access to electricity, and (b) could afford the machine, were going to benefit from the technology produced after two years of research.

An editorial note by Gibbon and Biggs to the article referred to in this study brings to light several important issues:

> For instance, it would appear that, if successful, the machines would only have benefited a minority of farmers, yet at the same time would have resulted in a displacement of labour. The criteria for the success of the threshing machine were not clearly laid down and seemed to be restricted to their economical and physical superiority over existing threshing methods. It is important that projects such as the one described in the above article should be prefaced by clear objectives on who the work is intended to benefit, the clear identification of client groups, a particular consideration of possible labour displacement effects and of equity issues.[70]

3. Bangladesh Rural Advancement Committee (BRAC)

Post harvest rice processing: Bangladesh

This study is taken from a report compiled by the Bangladesh Rural Advancement Committee (BRAC) which is a Bangladeshi voluntary organisation dedicated to the social and economic development of rural Bangladesh. BRAC is involved in a number of rural development projects in the area and its activities include the organisation of landless and marginal farmers, agriculture, fisheries development, youth organisation and training, health, family planning, nutrition, education, rural credit and employment generation. The emphasis that BRAC has placed on grass-roots experience has meant that their organisational abilities and policies have been directed by its observation. As an agency the last decade has taught them that there is a need for analysis and documentation of these experiences. Consequently they have a programme of training staff in the basic principles of observational research.

The following case study is one of a series of micro-studies undertaken in this scheme and attempts to demonstrate how a particular technology – in this case one with government backing – causes displacement of labour and disadvantages other groups in society.

Government policies of recent years have involved the spending of large amounts of capital on industrial plants. This includes rice mills – which by only employing a minimal number of workers effectively displace labour, thereby increasing unemployment and uneven income distribution. BRAC consider that the question of technology is one of vital importance.

At present 64-77 per cent of the paddy produced in the study area is husked in the household by women as is 90 per cent of the pre-milling processing (BRAC, 1979). However, two factors are causing a decline in these figures, (i) the spread of both custom and new major mills; and (ii) the financial remuneration that the women receive is lower than that necessary to support life.

Cheap electricity and low production costs for custom mills makes mechanical milling a very attractive proposition – but the benefits of such ventures are demonstrably limited to two groups. These are (a) the mill entrepreneurs; and (b) those surplus farmers who are able to dispense with female labour and take advantage of the cheaper facilities of the custom mills. Smallholder and marginal farmers, to whom labour is cheap and transport costs high, continue to use home husking methods. It is estimated that one mill displaces about 300 women, who all come from the very poorest classes, and it is estimated[71] that licensed custom mills are increasing at a rate of 5-7 per cent per annum:

> Taking the lower figure this gives an increase of 380 per year, displacing well over 1,000,000 women per year. It should be noted that the investment/output ratios of the technologies are almost the same, taking household technology as 100, custom mills are 136, but whereas investment in mills is expensive in foreign exchange in terms of raw materials, spare parts, and fuel, investment in local technology has largely been completed. There is a very large existing investment in *dhekis*, which have a very low rate of depreciation, which will be scrapped or highly under-utilised if custom mills capture the market.
>
> To summarise, subsidised mechanised mills are earning large profits for entrepreneurs and surplus farmers at the cost of poor women. The technology is profitable purely because it is labour displacing. Nothing new is created or produced, in fact the quality of *dheki*-husked rice is generally considered to be higher than that of machine-milled rice. While clearly a profitable activity for the entrepreneur, is mechanical milling a priority for the nation's scarce capital?[72]

Automatic rice mills have an even greater impact on labour as they remove the need for small-scale pre-milling processes. BRAC estimate that one automatic rice mill will displace 1,000 women and provide employment for only 18 skilled and 33 unskilled men. The only mill to have operated over a long enough period to formulate any kind of conclusion is the Comilla Modern Rice Mill. To 1978 it had only achieved a maximum output of 9 per cent of its capacity and was proving financially unusable.

On financial grounds alone there seems to be no justification to build any automatic mills. Custom mills though do seem to have a large cost advantage over household milling, in areas where transport is not a problem. However, they do displace the labour of a group to whom every increment of employment is important. BRAC, in an attempt to alleviate this problem, have suggested:

> An alternative strategy to the subsidised and unrestricted spread of custom mills owned by the rural rich would start by organising the existing household paddy-husking labour force

and through gradual increases in investment and improvements in technology improve their productivity and managerial capacity. IRDP and BRAC have already had considerable experience in organising and providing loans to paddy-husking cooperatives. By simply increasing working capital to enable the group to buy and sell larger quantities at the right time, profits can be substantially increased. In addition some small economies of scale in soaking, parboiling, drying have been achieved. It may be possible to introduce the simple steam parboiling device used in the custom mills, and either the more productive 'Teknaf' rotary husker or smaller-scale mechanical mills might be used. This process has already been started; it should not be impossible to develop it to the point where group members can acquire and run their own mill. There is no inherent reason why all the benefits of improved technology should be at the expense of, rather than in the interests of, the existing labour force.[72]

In summary, BRAC has identified three technologies which all affect poor rural women's employment prospects. It is suggested that government policies, often aimed at gaining foreign exchange, have been shortsighted and inappropriate to poverty groups' needs. BRAC has attempted to demonstrate that the right use of technological innovations can serve the purpose of efficiently utilising presently underused existing capital and labour:

> The two key factors are building and maintaining effective cooperative institutions at the local level and having a rational and employment oriented set of policies at the national level.[72]

C. INTERNATIONAL AGENCIES

1. World Bank, India

As a technology the development of efficient stoves has received both the attention and funding of a wide range of agencies – from small local groups to international agencies such as the World Bank.

Wood, representing the major fuel source for heating and cooking in developing countries, is the concern of rural women in particular. Responsible not only for the consumption of this resource, but also for its gathering and transportation, this group has felt the impact of the growing shortage of wood most deeply. It is through their allocation of resources that repercussions on other areas of life have been felt, e.g. family nutrition – as money previously allocated to food is diverted into the purchase of wood; soil fertility and crop yields – as dung and crop residues are collected and used as a wood substitute in domestic consumption.

Two areas of importance arise from these observations. The first is that in forestry projects women should be an obvious choice of client if they are to be the real beneficiary group. They will be the most likely to understand the implications of, and provide the local information for, proposed developments. Secondly, as those most intimately involved in food preparation, they and their methods of cooking must be the primary concern of any agency attempting to develop a stove.

The World Bank has been active in both these areas. For example, the:

> . . . World Bank Gujarat Project links wood production and consumption. An important aspect of this Project is the provision for communication units to be headed by female social workers whose task will be to determine and encourage the use of the most appropriate stove for the locality. The stove-builders assigned to each social worker will probably be men, even so, male artisans are usually more free to work with groups of women if a woman leader is supervising. Another interesting feature of this Project is its investigation into cooking behaviour, information which only women can supply. For instance, women accustomed to cooking in a squatting position are unlikely to accept an innovation such as a raised stove: the introduction of the female social workers will avoid such problems.[73]

A successful project has been recorded in Honduras, where the Lorena stove (developed in Guatemala) was introduced. One reason for this is likely to be that women were included in the Honduran group and trained in Guatemala in the construction of these stoves. These women were thus able to interact easily with local women as they possessed an innate understanding of the mechanics and problems of household affairs. In this particular case, credit was made easily available through a cooperative, and the economics of the technology were presented in a way that illustrated that a household could be equipped with a framed and roofed kitchen area and a stove at a monthly cost equalling the saving on fuel.

The significance of the lessons of (a) selecting the true beneficiary group; (b) learning with and through this group; and (c) developing technology locally, appear to have been completely misunderstood or ignored by the World Bank in other programmes. Notable exceptions in particular are those included in the Alternative Fuels Projects. In their Burundi Peat Project almost complete rejection of this fuel had been recorded. The simple explanation for this is that peat produces extremely heavy smoke, and houses in Burundi have no windows. In Upper Volta, charcoal stoves have not been adopted as (a) they are too expensive, and (b) they are not suited to the traditional methods of food preparation. One unfortunate offshoot from this concern with charcoal was that wood was used for charcoal making. This means that rural families are extremely short of fuel, as they no longer have access to wood and are unable to afford to buy charcoal. The stoves were developed in an attempt to increase the efficiency of charcoal burning, thus reducing the cost of this resource to the household. Moreover, there appears to have been little liaison with those who were actually responsible for the consumption or use of fuel as the non-adoption of these stoves illustrates.

What comes out of this discussion of stove projects is that women (the clients) have a vital role to play in designing, building and evaluating stoves. Although the stove design is often the focal point of a programme, the fuel issue must also be addressed in the light of women's interests and food preparation habits. Because of their familiarity with the issues involved, members of this group must also become the clients of agencies, as they will be the ones able most effectively to disseminate information regarding this technology.

2. ITDG (Intermediate Technology Development Group), Sri Lanka

The ITDG stoves project is almost the antithesis of the World Bank's attempts in this field and their projects continue to progress in Sri Lanka, Indonesia, Kenya and Nepal. The strength of these programmes lies in several issues. As an institution they have formed a very clear picture of their overall objectives and have formulated a definite strategy of development and information diffusion, Two institutional roles are played, (i) the hardware involving baseline technological research and information dissemination, and (ii) the software; acting as a support to, and providing links with and between, other AT agencies.

ITDG's design strategy has six stages:

(1) Initial survey of cooking practice and fuel usage (needs assessment).

(2) Field testing of existing stoves.

(3) Assessment of alternative stove designs.

(4) Laboratory testing and design modifications.

(5) Field testing and evaluation of new or modified stove designs.

(6) Extension/dissemination (including commercial manufacture).[74]

Evaluation studies are part of the programme, and all stages are carried out in collaboration with overseas groups (Dian Desa, Indonesia; Sarvodaya Shramadana Movement (SSM), Sri Lanka; Research Centre for Applied Science and Technology (RECAST), Nepal; Ministry of Energy, Kenya).

The Sri Lankan project has concentrated in the past year on strengthening SSM's technical abilities, and has helped them to form links with the Ceylon Institute of Scientific and Industrial Research (CISIR). In line with the above six stages there has been a refinement of the mud stoves already introduced in the Kandy district. These improvements have been aimed at improving the stoves' efficiency and their acceptability to users. Although charcoal is not an established fuel for cooking purposes the availability of this resource through the clearing of land on the Mahaveli Development Scheme has instigated government interest in its use – particularly for urban households. The CISIR stove has proved to be both efficient and acceptable to the client group. Savings of up to 30 per cent in fuel wood consumption have been recorded, and although this figure could be increased, the balance between acceptability, cost of production and efficiency dictates that at the present stage this is the 'happy medium'. Development work has been carried out on ceramic liners for stoves – the production of which has provided employment for two local potters; and, as a result of the evaluation study, training courses for extension workers have been implemented, and emphasis is being placed on increasing contacts with intended beneficiaries through village networks.

Similar positive results are recorded in the other projects. A strong and useful link has been set up with the Beijer Institute of the Netherlands, who in conjunction with the Technical Universities (Eindhoven) stoves group are hoping to establish a joint agroforestry/stove programme.

The reasons for the success of these projects (this is not to say that problems have not been experienced) are that the technical area, client and beneficiary groups (respectively, other agencies and stove users) were identified at the beginning of the programme. Through the use of local knowledge, local people and local institutions, research and development of the technology has been closely linked at all levels with the needs and resources available to those for whom they were designing. This concentration on field research has meant that there has always been an awareness of the multi-disciplinary nature of the environment, and provides an example of how it is possible to be conscious of this and use the knowledge gained without financing a large interdisciplinary research team, or wasting time in amassing large amounts of data that will never be used or be pertinent to the project in hand. Another significant feature of this project was that economists and other social scientists were drawn into the project to work with the engineers at critical stages of the work. The value attributed to monitoring and evaluation can be seen to be justified in that extension services were isolated as one weak area. Such studies mean that a programme is constantly subject to a scrutiny that can serve to change its direction, improve its performance and generally maintain a dynamic and realistic approach. An institution, accepting the value of this kind of study, must therefore be flexible enough to cope with what could perhaps be major changes in both theory and practice. It will be interesting to see the results of an assessment of the ITDG programme: an assessment which looks at the spread of improved wood stoves amongst the declared client group.

3. UNESCO/UNDP, Upper Volta

The following case study is one also where a process of programme evaluation was included as part of a project. Using a similar system to ITDG, needs were assessed, field testing of a suggested technology was introduced and its impact analysed. The positive results – although those desired were not achieved directly in the expected way – for the beneficiary group were assisted by the peripheral effects of the technology on another group.

Launched in 1967, the UNESCO/UNDP Women's Education Project (a non-formal programme based in Upper Volta) had two major objectives. These were, (i) to understand and document the reasons why women did not have full access to education, and (ii) to initiate experimental programmes to alleviate these problems so that women would be able to tackle the problems of poor health and low living standards.

Previous studies had pinpointed excessive work loads as a fundamental issue, thus three labour-saving technologies were introduced. The idea was that the use of the mechanical grain grinding mill, the cart for transportation of wood, water and produce, and the easily accessible new wells would give women enough time to take part in the educational programme – aimed at improving agricultural methods, literacy and health levels and generating income-boosting activities.

Under the programme, villagers selected village women and traditional midwives to attend training courses on knowledge dissemination. These women then organised the cooperative associations that were given access to the three technologies described above. Equipment was mainly for the use of the women belonging to the associations but could be rented out – the revenue from which was to be used on behalf of the crop.

In the subsequent evaluation exercise it was found that the grinding mills were not fully utilised. This was partly due to the fees for using the machine, and partly due to superficial mechanical problems, such as inadequate maintenance or parts not arriving or being too expensive to purchase. Carts, however, had a better record of adoption – their time-saving role was considerable, and there was also recorded an interesting shift in the distribution of labour. Men, previously unwilling to transport wood, water or crops on their heads, proved now to be quite amenable to carrying these commodities by cart. The effect of well-digging on reduction of work loads is unclear. Villagers dug the wells themselves, which yielded water for between one to eight months per year. During the dry season women still have to travel up to 5km to fetch muddy, worm-infested swamp water. However, the effects of the wells are felt in the related areas of adoption of health advice. This part of the project focused on water purification and sanitation disposal and has been seen to have positive effects.

In summary:

> The women themselves considered the technologies to be highly desirable; they saved time which could be spent on other activities. Despite the fact that the time saved tended to be used on other household tasks, women in the Project villages participated in education and income-earning activities in significantly greater numbers than the women in control villages. Activities directly relevant to daily life attracted the greatest participation.[75]

In terms of institutional lessons the contribution of this case study is in the area of problem identification. Having isolated poor health and low living standards as the objective, the programme sought to achieve the end of educating women on these subjects. The means were not seen simplisticly as only the provision of educational facilities, but, due to careful consideration of village conditions, were identified also as something to eliminate the obstacles that were inhibiting women's access to education. Two things here – firstly, the beneficiary groups were specifically identified; and secondly, the technology was introduced as an indirect but effective response to the objectives of the programme. This network of causes and effects is one therefore that should be traced as far back as possible. It is often unlikely that the root cause of the problem can be feasibly altered – but such a way of reviewing a given situation can lead to many apparently indirect or independent factors being taken into account.

4. AID (Agency for International Development), Nepal

Grain Storage in Nepal

This programme, whose objective was to reduce storage losses of grain by insects, rodents or moisture, provides an example of how western techniques have been adapted to local conditions.

It was reported that as much as 20 per cent of the Nepalese food grain harvest was lost during storage:

> Approximately 90 per cent of the population works on farms and the agricultural sector accounts for nearly 85 per cent of the country's export earnings. The provision of adequate storage facilities on farms thus ranks as a priority in efforts which are being made to improve agricultural marketing and to clear bottlenecks in Nepal's development.[76]

Consequently the Government of Nepal had an interest in the development of post-harvest storage techniques.

The pilot project experimented with several prototypes. Steel bins were shown to be uneconomical, and those constructed from mud and bamboo proved to need a high level of maintenance, on top of which they were neither rodent- nor insect-proof. After testing, the final prototype bin, made from reinforced concrete and utilising readily available materials and local labour, was taken to the Chitwan District of Central Nepal. Information leaflets describing the pros and cons and construction methods of the bins:

> ... assisted by AID field staff, Agricultural Development Bank of Nepal loan officers, local marketing cooperatives, and agricultural extension agents, each farmer paid for and supervised the building of his own site from start to finish,[76]

Twenty-seven farmers built their own bins – many of them adapting the innovation to their individual needs. Work was often carried out in cooperation with other farmers.

No data are available on the social or distributive aspects of this programme. However, it was noted that farmers had benefited in that they had learned to work together and had familiarised themselves with the services of local marketing cooperatives and agricultural extension agencies. They had also been made aware of the marketing problems associated with storage, credit and marketing demands. The fact that the technology reached the proposed client is attributable to the fact that this essentially technological-based project was developed locally, and therefore subject to the rigours of local conditions. The back-up services of credit and information also appear to have been good at establishing links and an easy rapport with farmers.

5. Wheat storage in Bangladesh

Finally a case of where a local Agricultural Ministry stopped itself wasting resources on storage technology. In this case the head of the Wheat Improvement Programme conducted a small survey (Ahmed n.d.) and found that rural women had utilised such items as cans and old polythene bags, and produced very low-cost but highly effective storage facilities for wheat seed. Although some experts were saying there were major storage problems, there was no need for concern in this area or for large amounts of funds to be put into research into improved wheat seed storage. Again, the lesson to be learnt is that the wheat scientific programme here had the interest, funds and flexibility to carry out a quick field survey of clients. Storage was found not to be one of their clients' problems.

V DRAUGHT ANIMAL TECHNOLOGIES

A. UNIVERSITIES AND RESEARCH ORGANISATIONS

ICRISAT (International Crops Research Institute for the Semi-Arid Tropics), India

The simple point made here is that large research institutes are trying to link their programmes. That this approach is very much in a formative stage is illustrated. The development of animal-drawn equipment has always been an important part of ICRISAT'S Farming Systems Research Programme. Since 1974 it has been the sole equipment system under investigation as tractors were seen to be a non-viable source of power to small farmers in the semi-arid tropics (SAT).

The objectives of the programme have been to reduce the energy (and time) to crop yield ratio. Much of the research at ICRISAT has been directed towards the ridge and furrow system of land management, and therefore the logical progression of the two technologies culminated in work being carried out to design a bullock-drawn machine suitable for this particular method of cultivation. The implement developed, drawn by two bullocks, has proved to be successful in some respects; however, its initial cost and that of the additional attachments is high for smallholder farmers. To enable these low-income farmers to take advantage of the broad-bed system, a locally designed wooden bullock cart has been modified by the addition of a tool bar, making it suitable for the aforementioned practices.

In their economic analysis presented at the SAT workshop, Binswanger et. al. (1979) conclude:

> . . . that all potential rental rates exceed the equivalent rental rate of traditional equipment at the factor price levels of Indian SAT farmers. These farmers, especially the small ones, cannot be expected to be interested in the package unless it provides them with sufficient yield gains per hectare to compensate for the added costs.

Whether or not this is feasible remains to be seen – but the attitude of those at present involved in tool-bar and animal draught research is that without other changes the only way smallholder farmers are going to benefit from their development is through cooperative ownership, or a 'fair pricing system' for the hire of such equipment. Perhaps the lessons to be learnt here then by the farming system research programme at ICRISAT are those of revaluation of (a) their criteria, or (b) their rather singular, technocratic approach to issues of technology generation and diffusion.

B. GOVERNMENT AGRICULTURAL DEPARTMENTS

Wum Area Development Authority (WADA), North West Cameroon

Interest began in increasing agricultural production in this region in 1965. The German development organisation involved attempted this by means of a training and settlement project, incorporating the use of tractors as the form of agricultural mechanisation. The failure of their project led to its being modified under the newly established Wum Area Development Authority in 1972. This case shows how experience gained during the course of a programme can be advantageously used if coordinators are willing to be flexible enough to incorporate changes to their original plans. It is also of interest because of the apparent impact of what was an increasingly disparate social situation as agriculture became more mechanised.

Three programmes were introduced – the Block Extension Programme, the Group Farming Programme and the Draught Animal Programme. The first initially continued to support farmers by subsidising tractors, but the project proved unsuccessful and only assisted the larger modern farmer. The second, offering short-term credits and guidance on rice cultivation to farmer-managed production groups, appears to have enjoyed a more positive outcome.

The Draught Animal Programme began in 1975 with the setting up of a training centre. The farmers trained are served by an extension network that provides one extension worker to 60 farmers. It is this group – local male farmers – who are considered to be the client group of the programme. They had no previous experience with draught oxen so adoption or non-adoption of the technology would be particularly noticeable. Traditionally cultivation was carried out by the women which meant that if animal traction were to be used the women would have to be 'directly included in the promotion measures as a client group'.[77] As this proved to be outside the terms of reference the programme has thus become closely linked with land settlement – where unused land reserves are used for settlement of young trained farmers. The resultant cost to the project of these measures is leading them to reconsider their defined client group and to extend their interests to influencing traditional cultivation methods

Evaluation of the use of draught animals in the Wum Region reveals that it is often farmers' experience and the length of time that they have been working with the animals that dictates the proportion of work still carried out manually:

> Data with any informative value cannot be anticipated until around two years after the farmers have been settled, since the ground must usually be tractor-ploughed in the first year of cultivation. Moreover, it cannot be assumed that the farmers will fully master the cultivation techniques until the third year. Initial experience has shown that during this year the cultivation area is expanded and additional tillage performed on a wage basis.[78]

It has been shown though that it is possible to increase agricultural income by up to 200 per cent by using animal draught. However, it is noted that many farmers in the area do not operate in a market-oriented way, thus profitability and wage labour is not assessed in monetary terms. The farmer is being forced into a cash economy though simply by the credit burden imposed upon him in the procurement of new equipment. Thus three changes are occurring as a result of this programme: (a) the entry into a cash economy, (b) the introduction of men into what was traditionally the female province of land using, and (c) farmers now have access to some form of mechanisation and therefore to produce higher yields and to become involved in the cash economy. The only ones previously benefiting from this method of production were the larger and more modern farmers.

The programme coordinators have recognised these three factors as is reflected in this statement:

> Project planning thus has the important task of creating the conditions for long-term and long-lasting changes by means of an appropriate innovation model.[79]

Here we have an agency willing to accept the responsibility of its innovation – and demonstrating this by responding to evaluation in a positive way and by being flexible enough in their approach to cope with a change from the initial direction, and an expansion from the technical focus, of their programme.

C. INTERNATIONAL AGENCIES

1. Overseas Development Administration (ODA), Botswana

Although this case is not one of the implementation of a technological programme it is one that shows why and how a development has taken place. It is also interesting to note that the animal-drawn tool has not been assessed on farmers' fields or in relation to anything other than their obvious and immediate physical environment.

The physical conditions of farmers in Botswana are those of low and variable rainfall (350-500mm/annum) which falls in a few heavy showers causing compaction of the shallow, weakly developed coarse-grained sands and sandy loams, and crust drying if a period of drying conditions occurs after the rains.

Hand cultivation for primary tillage is not normally practised, thus the 50 per cent of farmers who own no draught animals have to borrow them for ploughing and planting:

> Investment in the methods practised by the small farmer involves several radical changes and it is clear from previous extension experience and evidence from elsewhere that permanent change can be achieved by an improvement in the whole farming system.

The existing available animal-drawn equipment is inadequate and in many cases unsuitable for conditions in Botswana. Implements are of poor design, cover the gound slowly and encourage excessive losses of moisture. The results of field trials over the past three years indicate that cultivation implements such as chisels, sweeps, planters with press wheels and flint-bladed inter-row hoes are much more efficient than the current range of available implements, and are an essential prerequisite for the successful introduction of an improved crop production system.[80]

In answer to this the team have designed and constructed an animal-drawn two-wheeled implement carrier (the Versatool). All implements needed for ploughing, weeding, planting, inter-row hoeing and fertilising can be easily attached to the frame.

As no data are available on the rates of adoption or effects of this technology, it is left to Gibbon et. el. to assess future prospects:

> The median area of cleared land per farmer in SE Botswana is 9 ha. Most farmers have further areas of land allocated to them by the tribe.
>
> For an improved system using the Versatool, it is estimated that holding size would be limited to about 10 ha. by family labour availability. This size of unit can achieve an output that will provide subsistence for the farm family, cover all costs and, in most years, leave a cash surplus.
>
> When labour is not a problem, holding size will be limited by the capacity of the toolbar and the land available. A 14-ha. unit can provide a considerable cash surplus in good years. This is the most profitable system, the costs of equipment and draught power being spread over the maximum production area.
>
> Smaller farmers can benefit from these economies by sharing. Two six-ha. units could be worked with one set of equipment and draught animals. However, in the short-term the alternative for many farmers must be a system based on the single-furrow plough. A six-ha. unit, limited by the rate of effective planting, will provide subsistence most years with the occasional cash surplus.
>
> This alone is a considerable improvement over traditional

systems, where even in a good year only about 20% of farmers achieve subsistence.[81]

2. Canadian International Development Authority (CIDA), Sri Lanka

In collaboration with the Farm Mechanisation Research Centre (FMRC) CIDA has been investigating the possibilities of improving animal power used on the dry lands of Sri Lanka. This has particularly focused on the local meat cattle of the area as they are at present not fully utilised for farming operations.[82]

The group have developed a number of lightweight, reduced-draught implements and simple multi-purpose tool carriers and attachments which proved successful in field trials and experiments. However, the organisation of local manufacture of these items, field extension and economic evaluation have yet to be completed. What is of particular value in an evaluation of this programme is the reaction of rural youth to the techniques and hardware. About 60 per cent of the agricultural population lives by subsistence farming when most work is carried out manually. The conditions of manual work in agriculture are hard and do not apparently appeal to the majority of rural youth, who would rather chance their luck in the cities. However, the interest generated by the FMRC development of improved handtools has been further stimulated by the animal-powered technology, and it seems likely that such innovations may assist in persuading the rural youth to remain in farming. Although the hours of work may still be long, the physical drudgery has been reduced.

VI. LIVESTOCK

A. UNIVERSITIES AND RESEARCH ORGANISATIONS

Institute for Development Research (IDR), Bangalore, India

Cross breeding cattle programme

Although unsuccessful in benefiting its client group – the smallholder farmer – this programme does serve to illustrate why certain institutional recommendations were not adopted, and why certain constraints on production predetermined this.

IDR undertook a study of the cross-breeding of cattle programme, carried out under the Intensive Cattle Development Project, in an attempt to understand the problem. One of their initial findings was that cross-bred cattle were much more unevenly distributed among socio-economic groups than were the traditional milch animals. Small farmers and landless labourers tended to be dependent on this latter group of animals, whereas medium-size farmers used buffalo as their milk source. Although there was an overall awareness of the yield potential of cross-bred cows, it appeared that only the larger farmers were able to exploit this knowledge.

The scheme offered free artificial insemination for cattle to all farmers. However, demand for this service proved to be low. Analysis proved that it was the relationship between animal-draught demand, milk production and fodder availability that dictated the adoption rates of cross-bred cattle. It was generally recognised that male progeny from artificially inseminated cows were inferior for draught purposes. It also proved to be economically more feasible for small and middle farmers to breed for draught animals than for milkers.[83]

The larger farmers were able to take advantage of the milk yielding possibilities of cross-bred cattle as they had ample fodder resources. This conclusion is backed up the finding that, of the cross-bred cattle owned by small farmers, 75 per cent were not yielding milk due to the low level of their nutritional intake. Response to extension workers' suggestions that fodder grass be grown was low. This is immediately attributable to the fact that the fodder grass would have to be grown on irrigated land. Farmers felt that the value of the extra milk was low in comparison with that of the rice, sugar cane or vegetables normally cultivated on that land.

The findings of the IDR study are that the adoption of cross-breeding practices was low due to the constraints on small farmers. The project was attempted without any serious consideration of their needs and tried to introduce an economically non-feasible innovation. It is suggested that a concentration on the cross-breeding of buffalo might be more realistic. Buffalo management in areas of low-fodder resources is familiar to the smallholder farmer. This animal also requires lower initial financial input and is less costly to maintain than a cross-bred cow.

Perhaps the summary of this study is rather gloomy, but the conclusions can be constructive in terms of agency approaches to programmes. Firstly, it is obvious that, perhaps due to the scale of the whole Intensive Cattle Breeding Project, there was a lack of identification of specific local needs and conditions. This points to the value of employing small locally specific groups to analyse the real production situation. These could be another AT agency, or simply a local extension of the programme. What is important is that this type of research is done. Secondly, the extension workers were not noticeably successful in either disseminating information or helping in the adoption of the cross-bred cattle. This experience should serve to highlight the importance of this part of a programme and the need for its efficient delivery. Finally, the technological solution proposed was one that had been devised, and its implementation planned, in an alien environment by those unable to recognise the situation of the people who needed to be served with particular solutions in a specific way. Thus we can conclude that in terms of scale a large programme is unlikely to be successful unless it includes in it a component of locality-specific research, and is flexible enough to incorporate any findings of this. Ways in which to help the identified beneficiary group can only be assessed through this kind of approach, which must by definition be interdisciplinary in nature.

B. GOVERNMENT AGRICULTURAL DEPARTMENTS

1. Operation Flood – the Anand model, India

In 1949 Verghese Kurien went to Anand in the Kaira District of Gujarat to help the Indian Research Creamery set up a small plant making buffalo milk into powder. From this man's interest and idealism has developed Asia's most apparently successful rural cooperative. It is here that our interest lies – we hope to outline in the following brief description and analyse why and how this particular cooperative has been so successful.

In 1950 Kurien became manager of the newly created Anand cooperative. In technical terms he instigated four activities that have led to the production of healthy and improved yielding milch cattle. The first was the creation of a corps of qualified veterinarians who were both temporally and spatially available to all producers. The second was the establishment of hundreds of artificial insemination centres that could provide semen from high-quality bulls to farmers in the poorest and most remote villages. Thirdly, in the 1960s Anand was the motivating force in the production of a cheap and nutritional local cattle food. Finally, the breakthrough of finding a process of drying buffalo milk meant that seasonal gluts could be handled, thus ensuring that farmers always had a market. The Anand Milk Producers' Union Ltd. (AMUL) has further developed their processing capacity allowing production of other milk-based products such as sweets, cheese, baby food, ghee, butter, etc. They have also developed marketing facilities and networks that facilitate the distribution of their products (under the brand name 'Amul') throughout India.

The cooperative began in 1946 with two villages. It now

encompasses 831 villages. The reasons for the growth of this independent organisation are many:

> Farmers are paid twice daily, within 12 hours of the time they deliver milk to the societies' collection centres, and the centres are located directly in the villages where milk is being produced. All the cooperative societies are made up exclusively of villagers who supply milk and are run entirely by villagers elected from among the societies' members. The only other organisation in the cooperative is the Milk Producers' Union, which is headed by a Board of Directors elected by the elected heads of the village cooperative societies. At the Union level the Board of Directors hires professional personnel to maintain accounts, ensure quality control and efficiency in milk collection, manufacture and market milk products, and provide services for the cooperative's members. The services to members have mainly to do with the growth of healthy and productive cattle.[84]

Undoubtedly the commitment and organisational ability of Mr. Kurien has had much to do with the cooperative's success, but lessons can be learnt and principles drawn forth from the manifestation of this type of leadership. Initially there was an ideal of cooperative production aimed at achieving equity and efficiency for the poor of the area. By working with this group their problems became known. This village level familiarity meant that knowledge was gained in all areas of life, and thus the technical solutions that were developed were those that were geared to local needs and conditions. Local institutions were identified, and their capabilities recognised and strengthened. All producers were involved, they shared an ideal and a common identity which grew as the benefits of the project became obvious. Another factor of importance was that those responsible for the innovative thinking were willing to accept their responsibility. Operating on such a local level no other possibility was feasible.

Despite this apparent success closer scrutiny poses two questions, (i) have the interests of the very poor been, in fact, served by this programme? and (ii) is the production of milk a correct use of scarce resources? Although the general impact of the programme has been good, these two questions remain as yet unanswered – but the mere fact that they have been posed at all must indicate a recognition of these areas as problems. However, agencies can benefit from these experiences in that they emphasise the need for proper problem diagnosis on a multi-disciplinary level and in the direct contact with the beneficiary group. They must, if they do not have the institutional ability to work with the poor, ensure that the links between their interest group and the local institutions (whom hopefully they would be working through and with) are good and that a two-way flow of information is both unrestricted and unscreened. One final point, perhaps a by-product of this summary, is that in involving local institutions the knowledge gained in the implementation of a programme then remains in, and becomes part of, that locality's/country's experience, i.e. it is not exported with the foreign or larger agency.

The Anand model has served to act as a basis for the Indian Government's 'Operation Flood'. This profit-making scheme, initiated in 1971, hoped to have doubled India's milk production and established a nationwide grid of milk cooperatives for ten million farmers by the mid-1980s. In doing this the aims include establishing farmer-owned dairy co-ops in every city of more than 100,000 to assist the flow of urban money into rural hands. The programme has drawn directly on supplies made available by the World Food Programme (FAO/UN), and, in its recently launched second stage has made use of inputs from the EEC. The centre of the project is at Anand, where Kurien presides over the headquarters of the National Dairy Development Board (NDDB). NDDB teams are at present struggling to establish new dairy and cooperative societies.

The effects of attempting to transfer the Anand model are mixed. The smaller farmers have benefited to a certain degree such as in the introduction of cross-bred cattle, providing them with higher yielding and healthier cattle; and in the provision of veterinary services, which has lowered the mortality rate considerably. However, the questions raised concerning the effect of implementation of the Anand model on a small scale have become increasingly obvious at a larger scale. There are three points that are apparently in contradiction to Operation Flood's stated objectives of developing:

Animal husbandry as an important economic occupation for

the small and marginal farmers and agricultural labourers. As with other investments the commercial programmes of animal husbandry can be quite remunerative, there is an expectation for the richer classes to enter into these occupations to the detriment of the traditional (poorer) classes dependent on animal husbandry. It will be a major aim of the 5th Plan to ensure that the traditional classes get the lion's share of the new programmes of animal husbandry development.[85]

It has been noted (Crotty, 1980) that larger farmers are managing to obtain higher incomes per hectare than smallholder farmers – something attributed to the ability of the larger producers to sell fodder to the landless. However, it is also fair to state that despite this the landless have still benefited from the milk co-ops in that the income generated from milk production has increased of late. The second issue is one of cost. Operation Flood has raised the price of milk in rural areas – thereby reducing the availability of this source of protein to the rural poor. The main group at risk from this aspect of the programme are infants, as adults can partake of a nutritionally adequate diet without the inclusion of milk – indeed the extra income generated from the sale of milk could be used to purchase more basic foodstuffs. The third point is possibly the most important, and follows from the second. Milk production has become a more profitable occupation than grain and legume production for human consumption. In short, this latter activity has been curtailed by grain being diverted to cattle feed, and/or land being used for the growth of fodder crops. Unfortunately the scenario is apparently becoming one of the diversion of scarce resources to the production of a rather inefficient form of food, that can only be afforded by the richer sections of society. The rural poor are benefiting in terms of better quality and healthier cattle, and are receiving more regular and a higher income from their milk producing activities. However, disparities are beginning to show between the richer and poorer sections of the community; the question remains whether or not such large-scale milk production is a correct usage of scarce resources. There has also been some criticism levelled at the scheme as it is claimed that women, traditionally those making the decisions about dairying, are being ousted by the cooperative committees.

The interest in this juxtaposition of cases lies in that it demonstrates how difficult it is to apply a model derived from one area of experience to another. There are so many differences that have to be considered – not the least the question of scale. In the first example the scale was small and the research needs-oriented; in the second the scale of the project is national, and from what can be understood research into locality-specific needs and problems in small.

What appears to have happened is that there has been an attempt to apply the outward manifestation of a successful programme to another situation, rather than an attempt to draw out the underlying principles for that initial success and developing a programme upon them. In the case of Operation Flood it is those principles that are being denied by the very application of their expression in one situation to another. Therein lies the lesson that, what is important to planning and utilising experiences from elsewhere, is the correct evaluation of the subsisting reasons, not the transposition of the structural, technocratic and bureaucratic forms that were generated from them.

2. Bee-keeping cooperative, Honduras

This example is one of where the potential of an often unrecognised livestock enterprise has been exploited. The area itself is of interest as it illustrates how activities that are often not considered in terms of 'agriculture' can be capitalised upon to provide income and employment to certain groups.[86]

Bee-keeping and the marketing of products is a particularly useful occupation for several reasons. Firstly, the scale of the operation is variable – from a few domestic hives to a much larger operation, thus the labour commitment can be that suited to the producer's situation. Secondly, it does not compete with or for other agricultural resources. Thirdly, the financial inputs are minimal. Finally, the primary product, honey, can be used as a cash crop or as an important nutritional input to the family household. The secondary product, wax, can provide the raw material for household industries such as candle-making.

As on-farm employment, bee-keeping is proving attractive to rural women, often already involved in the production of crops for household consumption. In Honduras the workload has been distributed amongst this particular group by the formation of cooperatives:

> The cooperative was established with advice from the Honduras Ministry of Natural Resources in 1981, and with a bank loan which must be repaid within four to five years. The cooperative of 18 women and two boys receive regular advice from a Ministry project co-ordinator. They were initially given a training course in bee-keeping at their village by the government education authority. With the money from the bank loan, a honey-extracting house (including water on tap), and 34 colonies in Langstroth hives, were established. The members divided into three groups which carry out in turn the necessary management for a ten-day period, paying themselves the equivalent of the average (male) daily wage for their labour. The Ministry project co-ordinator gives assistance to each group, especially advice on finances. The only written records kept at the moment, by the elected secretary, are the financial accounts. The extracted honey is sold locally, and the remaining comb honey is used by the women themselves. The beeswax is recycled by making it into a foundation which is given to the bees to draw out into comb for honey storage. The first year's honey harvest was very promising, well above the national average, and allowed the women to pay back a large proportion of the bank loan.[87]

The project appears to be functional in that it is providing the predominantly female members of the co-op with a source of income. No data are available regarding the effects of this, or how the women have incorporated this extra activity into their already busy working lives. However, the fact that the first year's harvest was both financially and quantitatively successful indicates that the women are not finding this added activity too burdensome.

Structurally the Ministry of Natural Resources has placed emphasis on developing the strength of a local institution and has been innovative (and willing to take the financial responsibility) in the area it has looked at.

VII. SUMMARY AND CONCLUSIONS

A. DEFINITION OF CLIENTS

In general, it was found that when a local agency has a genuine commitment to a policy objective of addressing the problems of a specific group of poor people they have not had difficulty in adequately defining, in a pragmatic way, their poor clients. However, it is important that local agencies not only identify their client beneficiary group, but that they also in the course of the programme evaluate the implications of their work, not only on the defined beneficiary group, but also on other local poverty groups. Thus the monitoring of direct and peripheral effects on poverty should be a permanent feature of a programme.

B. CHARACTERISTICS OF VIABLE POVERTY-FOCUSED LOCAL AGENCIES

There appears to be no immediate pattern as regards whether specific types of government agency or any NGO programme are, in some sense, better or worse in helping to generate and diffuse technology. Characteristics which run across viable public sector and non-government organisations are:

- A genuine commitment to identifying specific poor client groups and to monitoring the effects of their activities against poverty criteria;
- A high professional ability to be able to diagnose technical problems and be able to change, sometimes radically, research

and extension priorities as more information from their client group and research is assimilated;

- A continuous commitment to household, farm and village level analysis and to seeing this information as a major guiding force in the agency's activities;
- An explicit recognition, and pride in the fact, that the agency has had to change, at least some of, its priorities and programmes as it recognises that it had gone about its work in an unproductive way – at least as far as the poor were concerned;
- A commitment to using, where relevant, outside knowledge to help address local problems but recognising that outside knowledge on hardware and software was a relatively minor input compared with a local ability to select, modify and use this information in addressing local problems;
- An ability to use and act on the results of multidisciplinary analysis of client group's problems;
- In the context of addressing poverty problems, a recognition that effective communications between scientists within and between disciplines, between institutions, between governments and NGOs, between aid agencies, etc., are extremely important and difficult problems to address;
- A recognition of the inherent problems of 'special projects'. The strengthening of a local R and D and extension capability is often given as a priority for the work of any agency. However, special projects with special funds, special technologies, special research methods, special staff may do more to undermine existing informal and formal research and extension capabilities than to strengthen them;
- A recognition of the full depth of the political, cultural and institutional dimensions which are involved when applying a simple 'problem solving' approach to technology issues. The most viable projects are those with personnel that know that the application of a problem solving approach is easier said than done!
- An acceptance of technology responsibility. One of the important lessons from the case studies was, in some cases, a recognition by those developing and diffusing a technology that they were being highly selective as regards the choice of

techniques, practices and methods which they were identifying, developing and presenting. In this they were taking responsibility for what they were doing. As an example, some of the most viable agencies were the ones which described the 'inappropriate' technologies and methods they had started with and the types of mistakes they had made. The idea that agencies are able to demonstrate to poor clients a full range of viable alternatives – from which the poor select those which are most suitable to their need – is symptomatic of an approach which is unrealistic and irresponsible. Some of the situations where social scientists and agriculturalists have worked most productively together have been where it was clear that both groups would be held responsible for the adoption or non-adoption of technology amongst the pre-identified client groups. The skills of the social scientists and economists helped not only in diagnosing problems but also in forecasting the socio-economic circumstances which poor people might expect to face in the future;

– A recognition of the need to use practical methods which ensure that poor client groups are the main beneficiaries. In the review it was seen that it is difficult to ensure that R and D and extension programmes concentrate, and remain focused, on the problems of the rural poor. Practical methods for increasing the chances that poor client groups are the main beneficiaries of agricultural technology programmes involve the use of checklists and decision matrices.[88,89,90,91]

C. POLICY ISSUES

Finally, throughout the case studies it was clear that local agencies were affected a great deal by government policy. This was apparent in two major respects. One is that promotions and salaries in research and extension agencies are not normally based upon the diffusion of technology amongst poor clients or on attempts to address poverty problems in a multidisciplinary framework. Indeed, one of the reasons for the growth of the 'appropriate' technology movement is a response to this disparate nature of rewards and incentives within conventional research systems. The implicit bias in science and technology policy – which is reflected by the existing institutional structures and reward systems – will need to be addressed if there is to be a widespread and long-term application of science and technology to poverty issues.

The other factor is the very obvious role of government economic and social policy in determining whether specific technologies can be used by or will benefit poor people in different situations. There are many situations in which technology could be used to benefit the poor if government pricing, credit and other policies were changed. Since the focus of this paper is on issues concerned with the internal organisation and management of scientific systems, it has made no attempt to look at this issue. It is, however, one which is of critical importance and should rightly form the basis of a separate review.

NOTES

1. Zandstra.
2. Kenya, 1979.
3. Gilbert et al., 1980, pp.2-3.
4. Note that CIMMYT's mandate restricts the focus of its research to maize, wheat, triticale and barley.
5. Collinson, 1982, pp.29-32.
6. Collinson, 1982, p.33.
7. Collinson, 1982, pp.40-41.
8. Byerlee et al., 1982.
9. Winkelmann, 1976.
10. CIMMYT, 1974, p.79.
11. See CIMMYT Information Bulletin 27 'From Agronomic Data to Farmer Recommendations' (Perrin et al., 1979).
12. Brammer, 1978, p.3.
13. Perrin and Winkelmann, 1978, p.893.
14. IRRI's programmes under the Component Technology Criteria include research on weeds in dryland crops planted after rice; effects of crop rotation on weed growth; rice stubble management and cowpea insects; establishment of corn after rice; soybeans after wetland rice; and variety testing for cropping systems.
15. Gilbert, 1980, p.90.
16. Binswanger, 1977.
17. Bantilan, 1974.
18. Bantilan, 1974.
19. Binswanger, 1977, p.57.
20. Gibbon, 1981.
21. Gibbon, 1981, p.14.
22. Gibbon and Martin, 1978, p.23.
23. IITA, 1979, p.65.
24. Zandstra, 1979, p.9.
25. Zandstra, 1979, p.292.

26. Morss, 1976, p.295.
27. Morss, 1976, p.307.
28. Zandstra, 1979, p.12.
29. Biggs, 1982a.
30. Palmer, 1975, p.55.
31. Morss, 1976, p.44.
32. Morss, 1976, pp.45-46.
33. Corven, 1982.
34. Morss, 1976, p.24.
35. Dey, 1980.
36. A detailed analysis of 'formal' and 'informal' R and D and the relationship between the two is given by Biggs and Clay (1981).
37. Biggs, 1980.
38. Quasim, 1981.
39. Yunus, 1978.
40. Yunus, 1981.
41. Edwards et al., 1978.
42. 60,000 MOSTI (Manually Operated Systems of Tubewell Irrigation) and 20-30 thousand dug wells (Islam, 1976).
43. Thomas et al., 1976, pp.461-462.
44. Wood, 1982.
45. Wood, 1982, p.46.
46. Gachukia, 1982, p.20.
47. O'Kelly, 1982.
48. '. . . where water is pumped from a pond or reservoir to the top of a series of adjoining hills. It then flows into the higher fields and from there gradually percolates into lower fields until it collects in a reservoir below. It is then pumped to the higher fields again.' (Franda, 1983.)
49. Franda, 1983, Fig. 3.
50. Byrne, 1983.
51. Clay, 1980.
52. Appu, 1974.
53. Greeley, 1982, 1982a.
54. Harriss, 1979.
55. Timmer, 1974.
56. Maxwell, 1982, p.24.
57. Maxwell, 1980(b).

58. Maxwell, 1982(a).
59. Maxwell, 1982(a), p.26.
60. Coursey, 1982.
61. Coursey, 1982, pp.12-13.
62. Coursey, 1982, p.16.
63. CIP, 1980.
64. Rhoades and Booth, 1982.
65. Such as 'Farmer inputs turn direction of agricultural research' (CIP, 1981) and 'Farmers help scientists change research objectives' (CIP, 1981a).
66. Lawrence-Jones, 1982.
67. Lawrence-Jones, 1982, p.29.
68. Rhoades and Booth, 1982.
69. Kennedy, 1979.
70. Bose and Infield, 1982.
71. Harriss, 1979.
72. BRAC, 1979.
73. Scott, 1982, p.13.
74. ITDG, 1982.
75. McSweeney, 1982.
76. Wilson, 1974.
77. Wagner and Munzinger, 1982.
78. Wagner and Munzinger, 1982, p.397.
79. Wagner and Munzinger, 1982, p.400.
80. Gibbon et al., 1974, p.229.
81. Gibbon et al., 1974, p.234.
82. Pillainayagem, 1982.
83. Rajapurohit, 1979.
84. Franda, 1979.
85. 'Draft Fifth Five Year Plan', para. 1.202, India Planning Commission, 1974.
86. Re FUNDAEC where fish and poultry were used as the basis for improvement of conditions for, in particular, women who had no access to other production activities.
87. Nixon, 1982, p.12.
88. Biggs, 1978.
89. Biggs, 1981.
90. Chambers, 1978.
91. Carruthers and Clayton, 1977.

BIBLIOGRAPHY

Adams, J. M. and Shulter, G. G. M., 1978. 'Losses caused by insects, mites and micro-organisms' in Harriss, K. L. and Lindblad (eds.) *Post-Harvest Grain Loss Assessment Methods.* American Association of Cereal Chemists.

Ahmed, S. M., n.d. 'A report on survey of wheat seeds stored by the farmers for sowing in 1976-77', mimeo report, Bangladesh Agricultural Research Institute.

Appu, P. S., 1974. 'The bamboo tubewell: a low cost device for exploiting ground water.' *Economic and Political Weekly,* Vol. 9 (26), 29 June.

Bantilan, R. T. and Harwood, R. R., 1974. 'Weed management in multiple cropping systems.' Paper presented at the General Meeting of the Weed Science Society of the Philippines, 6 December 1974.

Bell, C. L. G. and Duloy, J. H., 1974. 'Rural Target Groups' in Cheney, H., Ahluwalia, M. S., Bell, C. L. G., Duloy, J. H. and Bell, R. *Redistribution with Growth,* 1974.

Biggs, S. D., 1978. 'Planning rural technologies in the context of social structures and reward systems.' *Journal of Agricultural Economics,* Vol. XXII (3), pp.267-274.

Biggs, S. D., 1980. 'Informal research and development: farmer resistance to new technology is not always a sign of backwardness.' *Ceres,* No. 76, Vol. 13 (4), pp.23-26.

Biggs, S. D., 1981. *An Evaluation of Appropriate Technology,* Report No. EV 262, Overseas Development Administration, London.

Biggs, S. D. and Clay, E. J., 1981. 'Sources of innovation in agricultural technology.' *World Development,* Vol. 9 (4), April, pp.321-336.

Biggs, S. D., 1982(a). 'Generating agricultural technology: triticale for the Himalayan Hills.' *Food Policy,* Vol. 7 (1), February, pp.69-82.

Biggs, S. D., 1982(b). 'Monitoring and control in agricultural research systems: maize in Northern India.' *Research Policy*, Vol. 12 (1), pp.37-59.

Biggs, S. D., 1982(c). 'Agricultural research: a review of social science analysis.' *Discussion Paper* 115, School of Development Studies, University of East Anglia, Norwich, UK.

Binswanger, H. P. and Shetty, S. V. R., 1977. 'Economic aspects of weed control in semi-arid tropical areas of India.' Proc. Weed Soil Conference and Workshop in India, 1977, ICRISAT, India.

Binswanger, H. P. and Ryan, J. G., 1979. 'Studies as a focus for research and technology adoption.' Paper presented at International Symposium on Development and Transfer of Technology for Rainfed Agriculture and the SAT Farmer. August 1979. Hyderabad, ICRISAT.

Binswanger, H. P., Ghodate, R. D. and Thierstein, G. E., 1979. 'Observations on the economics of tractor, bullock and wheeled tool carriers in the semi-arid tropics of India.' Paper presented at the International Workshop on Socio-economic Constraints to Development of Semi-Arid Tropical Agriculture, 19-23, February 1979. Hyderabad, India.

Booth, R. H., 1974. 'Post harvest deterioration of tropical root crops.' *Tropical Science*, Vol. 16, No. 1, pp.49-63.

Booth, R. H. and Coursey, D. G., 1974. 'Storage of cassava roots and related post-harvest problems.' *Cassava Processing and Storage, IDRC Monograph*, IDRC-031e, International Development Research Centre, Ottawa.

Bose, A. R. and Infield, J. A., 1982. 'Experiences in the design of a manually operated wheat threshing machine.' *Appropriate Technology*, Vol. 9, No. 1, June 1982, pp.11-13.

Bose, S., 1974. 'The Comilla cooperative approach and the prospects for broad-based green revolution in Bangladesh.' *World Development*, Vol. 2, No. 8, August 1974, pp.21-28.

BRAC, 1979. *Appropriate Technologies Under Pressure – Cases of the Cotton Textile Industry and Post Harvest Rice Processing.* Dacca.

Brammer, H., 1978. *Wheat Strategy Considerations 1978-1985.* Soil Survey Interpretation Division. Directorate of Soil Survey, Dacca, 15 July, 1978.

Brammer, H., 1979. 'Learn from the best farmers.' Ministry of Agriculture and Forests, Dacca, January.

Byerlee, Derek, Harrington, L. and Winkelmann, D. L., 1982. 'Farming systems research: issues in research strategy and technology design.' *American Journal of Agricultural Economics*, Vol. 64 (5), December, pp.897-904.

Byrne, H., 1983. 'Clean water in Kola.' *Waterlines,* Vol. 1, No. 3, January 1983, pp.21-23.

Carruthers, I. D. and Clayton, E. S., 1977. 'Ex-post evaluation of agricultural projects: its implication for planning.' *Journal of Agricultural Economics*, Vol. 28 (3), September, pp.305-318.

Chambers, R., 1978. *Project Selection for Rural Development: Simple is optimal world development,* Vol. 6, pp.209-219.

Chenery, H., Ahluwalia, Bell, C. L. G., Duloy, J. H. and Jolly, R., 1974. *Redistribution with Growth*, Oxford University Press, e1974.

CIMMYT, 1969. *The Puebla Project 1967-69. Progress report of a programme to rapidly increase corn yields on smallholdings.* CIMMYT, Mexico.

CIMMYT, 1974. *The Puebla Project: Seven Years of Experience 1967-73. Analysis of a programme to assist small subsistence farmers to increase crop production in a rainfed area of Mexico.* CIMMYT, Mexico.

CIP, 1980. Centro Internacional de la Papa, 1980. *Profile.* CIP, Lima, Peru.

CIP, 1981. 'Farmer inputs turn direction of agricultural research.' Centro Internacional de la Papa, *Circular*, Vol. IX (3), March.

CIP, 1981(a). 'Farmers help scientists change research objectives.' Centro Internacional de la Papa, *Circular*, Vol. IX (3), April 1981.

Clay, E. J., 1980. 'The economics of the bamboo tubewells: dispelling some myths about appropriate technology.' *Ceres*, No. 75, Vol. 13 (2), May/June, pp.43-47.

Clay, E. and Schaffer, B. (eds.) (forthcoming). *Room for Manoeuvre in Public Policy*, Heinemann.

Collinson, H. P., 1982. 'Farming Systems Research in Eastern Africa: the experience of CIMMYT and some national agricultural research services, 1976-81.' *MSU International*

Development Paper No. 3, Department of Agricultural Economics, Michigan State University, East Lansing, Michigan, USA.

Corven, J., 1982. 'Ideas from FUNDAEC. Integrated farming in Colombia.' *VITA News*. January 1982.

Coursey, D. G., 1982. 'Traditional tropical root crop technology: some interactions with modern science', in *IDS Bulletin*, Vol. 13 (3), pp.12-20.

Crotty, R., 1980. *Cattle, Economics and Development*. Commonwealth Agricultural Bureaux.

Dey, J., 1980. 'The Socio-economic organisation of farming in The Gambia and its relevance for agricultural development planning.' *Agricultural Administration Network Paper*, No. 7, ODI, London 1980.

Edwards, C., Biggs, S. D. and Griffith, J., 1978. 'Irrigation in Bangladesh: on contradictions and underutilised potential.' *Development Studies Discussion Paper*, No. 22, University of East Anglia, Norwich, U.K.

Ford Foundation, 1982. 'Crops Research in West Africa.' *Ford Foundation Letter*, December, 1982, p.5.

Franda, M., 1979. *Small is Politics: Organisation Alternatives in India's Rural Development*. Wiley Eastern Ltd., New Delhi.

Franda, M., 1983. 'Fair share irrigation for India's poor.' *Ford Foundation Letter*, Vol. 14 (1), 1 February, 1983.

Fryer, J., 1981. 'Benefits from the Indian sacred cow.' *The Geographical Quarterly*, July 1981, pp.617-622.

Gachukia, E., 1982. Women's self-help efforts for water supply in Kenya. *Appropriate Technology*, Vol. 9, No. 3, December 1982, pp.19-21.

Gaikwad, V. R., Desai, G. M., Mampilly, P. and Viyas, V. S., 1977. *Development of Intensive Agriculture: Lessons from IRDP*. Indian Institute of Management, Ahmedabad, India.

Gibbon, D., Harvey, J. and Hubbard, K., 1974. 'A minimum tillage system for Botswana.' Development Reprint, No. 5. *World Crops*, Vol. 26, No. 5, September/October 1974, pp.229-234.

Gibbon, D. and Martin, A., 1978. 'Food legumes in the farming system: A case study from Northern Syria.' *Development Studies Reprint*, No. 94.

Gibbon, D., 1981. 'A systems approach to resource management and development in the ECWA region: the ICARDA experience 1977-80.' *Discussion Paper* No. 104, School of Development Studies, University of East Anglia, Norwich, U.K.

Gilbert, E. H., Norman, D. W. and Winch, F. E., 1980. 'Farming Systems Research: a critical appraisal.' *MSU Rural Development Paper*, No. 6, Department of Agricultural Economics, Michigan State University.

Greeley, M., 1982 (a). 'Farm level post-harvest food losses: the myth of the soft third option.' *IDS Bulletin*, Vol. 13 (3), June, pp.51-60.

Greeley, M., (ed.) 1982. 'Feeding the hungry: a role for post-harvest technology.' *IDS Bulletin*, Vol. 13 (3).

Hammerton, J. L., 1974. 'Problem of herbicides use in peasant farming.' Paper presented at the annual meeting of Weed Science Society of America. 1974, mimeo.

Harriss, B., 1979. 'Post-harvest rice processing systems in rural Bangladesh: Technology, economics and employment.' *Bangladesh Journal of Agricultural Economics*, Vol. II (1), June, pp.23-50.

IITA, 1979. *Annual Report for 1978*, Ibadan, Nigeria.

IRRI, 1977. *Constraints to High Yields on Asian Rice Farms: an Interim Report*, Los Baños, Philippines.

IRRI, 1978. 'Cropping systems research highlights.' Mimeo. Los Baños.

Islam, T. R., 1976. 'Economics of hand tubewell irrigation at farm level in Bangladesh.' Paper presented at the National Seminar on Water Management and Control at the Farm Level. 15-20 March, 1976. Bangladesh University of Engineering and Technology, Dacca.

ITDG, 1982. *Stoves Project Annual Report* (September 1981-August 1982). ODA Research Scheme No. R3598. ITDG November 1982.

Kampen, J., 1979. 'Farming Systems Research and technology for the semi-arid tropics.' Paper presented at the International Symposium on Development and Transfer of Technology for Rainfed and the SAT Farmer, August, 1979. ICRISAT, Hyderabad, India.

Kennedy, J., 1979. 'Making manioc starch in Sri Lanka: a rural industrial enterprise.' *Appropriate Technology*, Vol. 6, No. 3, pp.4-6.

Kenya 4th Five-year Development Plan, 1979. Government of Kenya, Nairobi.

Ker, A. D., 1979. *Food or Famine: an account of the crop science program supported by the International Development Research Centre*, IDRC – 143e International Development Research Centre, Ottawa.

Krantz, B. A., Kampen, J. and Associates, 1976. *Annual Report of the Farming Systems Research Programme*. ICRISAT, Hyderabad, India.

Lawrence-Jones, W., 1982. 'The development, diffusion and consequences of technology: a case study of seed potato storage.' Overseas Development Group, University of East Anglia, Norwich.

Martin, A., 'Concepts and approaches to Farm Systems Research in Syria 1977-80.' *Discussion Paper* No. 120. School of Development Studies, University of East Anglia, Norwich, U.K.

Maxwell, S., 1980. 'Differentiation in the colonies of Santa Cruz: causes and effects.' *Working Paper No. 13*, CIAT, Santa Cruz, Bolivia.

Maxwell, S., 1982. 'Harvest and post-harvest issues in Farming Systems Research.' Institute of Development Studies. *Bulletin*, Vol. 13 (3), pp.21-32.

McSweeney, B., 1982. 'Time to learn for women in Upper Volta.' *Appropriate Technology*, Vol. 9, No. 3, December 1982, pp.27-30.

Morss, E. R., Hatch, J. K., Mickelwait, D. R. and Sweet, C. F., 1976. *Strategies for Small Farmer Development*, Vol. I, Vol. II: Case Studies. Westview Press.

Nixon, M., 1982. 'Women and beekeeping.' *Appropriate Technology*, Vol. 9, No. 3.

Ogborn, J., 1969. 'The potential use of herbicides in tropical peasant agriculture.' *Pans*, 15: 9-11.

O'Kelly, E., 1982. 'Women's organisations: their vital role in development.' *Alternative Technology*, Vol. 19, No. 3, December 1982.

Palmer, I., 1975. 'Problems of access: Indonesia and Malaysia.' *Development and Change*, Vol. 6, No. 2, April 1975.

Pant, N., 1982. 'Close look at SFDA's.' *EPW*, XVII (6), February 1982.

Perrin, R. K. and Winkelmann, D., 1976. 'Impediments to technical progress on small versus large farms.' *American Journal of Agricultural Economics*, Vol. 58 (5), December 1976, pp.888-894.

Perrin, R.K., Winkelmann, D. L., Moscardi, E. R. and Anderson, J. R., 1976. *From Agronomic Data to Farmer Recommendations: An Economic Training Manual.* CIMMYT, Mexico.

Pillainayagem, M. G., 1982. 'Research into farm power and equipment in Sri Lanka.' Paper presented at the Regional Seminar on Farm Power, 25-29 October 1982, at the Agrarian Research and Training Institute, Colombo, Sri Lanka.

Quasim, M. A., Saha, S. R. and Saha, B., 1981. A study on the 'impact of Grameen Bank Project operation on landless women', Bangladesh Institute of Bank Management, Dhaka.

Rahim, S. A., 1972. 'Rural cooperatives and economic development of subsistence agriculture.' BARD, Comilla 1972 (mimeo).

Rajapurohit, A. R., 1979. 'Cross-breeding of Indian cattle: An evaluation.' *Economic and Political Weekly*, Vol. XIV, Nos. 12 and 13, 24-31 March 1979, pp.119-128.

Redclift, M. 'Production programmes for small farmers: Plan Puebla as myth and reality.' *Economic Development and Cultural Change*, forthcoming.

Rhoades, R. E. and Booth, R. H., 1982. 'Farmer-back-to-farmer: a model for generating acceptable agricultural technology.' *Agricultural Administration*, Vol. 11, pp.127-137.

Scott, G., 1982. 'Forestry projects: How woman can help – and help themselves.' *Appropriate Technology*, Vol. 9, No. 3.

Srivastava, J. P., 1973. 'Prospect of triticale as a commercial crop in India.' 'Triticale: Proceedings of International Symposium,' El Batan, Mexico, 1-3 September 1983. *IDRC Monograph*, No. 021e, 1974. International Development Research Council, Ottawa.

Technical Advisory Committee, 1978. *Farming Systems Research at the International Agricultural Research Centres*,

Consultative Group on International Agricultural Research, Washington.
Thomas, J. W., Shahid, J. Burki, Davies, D. G. and Hood, R. M., 1976. 'Public works programmes: Goals, results, administration' in: Hunter, G., Bunting, A. H. and Bothall, A. (1978) *Policy and Practice in Rural Development.* Croom Helm, London.
Timmer, C. P., 1974. 'Choice of technique in rice milling on Java,' Research and Training Network Reprint, Agricultural Development Council Inc., New York. (See also the comments and discussion which follows the article.)
Wagner, C. M. and Munzinger, P., 1982. 'Introduction of draught animals in North-West Cameroon by the "Wum Area Development Authority"' in *Animal Traction in Africa*, Munzinger, P., 1982, Eschborn.
Werge, R., 1977. *Potato Storage Systems in the Mantaro Valley region of Peru,* The British Centre, Lima, Peru.
Whitcombe, R. and Carr. M., 1982. *Appropriate Technology Institutions: A Review.* Occasional Paper No. 7. Intermediate Technology, London.
Wilson, T., 1974. 'Reducing Nepal's grain losses.' *War on Hunger*, Vol. VIII, No. 3, March 1974.
Winkelman, D., 1976. 'The adoption of new maize technology in Plan Puebla, Mexico.' CIMMYT, Mexico.
Wood, G. D., 1982. 'The socialisation of minor irrigation in Bangladesh – a review and financial analysis of the 1981-82 season for 83 groups supported by PROSHIKA as part of the "Irrigation Assets for the Rural Landless".' Action Research Programme. PROSHIKA, Dhaka.
Yunus, M., 1978. 'Bhunisheen Samiti (Landless Association) and Mohila Samiti (Women Association) in Jobra and other villages.' Paper presented at the National Seminar on Rural Development, Dacca, 24-29 April 1978.
Yunus, M., 1980. 'Grameen Bank Project: towards self-reliance for the poor.' Paper presented at the Colloqium of Bankers on Providing Credit Facilities for Rural Women, 1-5 December 1980, Tangail, Bangladesh.

Yunus, M., 1981. 'Rural/agricultural credit operations in Bangladesh.' Paper presented at the Annual Conference of the Bangladesh Economic Association, Dacca, 30 April 1981.

Zandstra, H., Swanberg, K., Zulberti, C. and Nestel, B., 1979. *Caqueza: Living Rural Development.* International Development Research Centre (IDRC), Ottawa, Canada.

www.ingramcontent.com/pod-product-compliance
Lightning Source LLC
Jackson TN
JSHW011411130125
77033JS00024B/963

* 9 7 8 1 8 5 3 3 9 3 6 5 5 *